Writing successfully in science

Writing successfully in science

MAEVE O'CONNOR

Secretary-Treasurer,
European Association of Science Editors

With cartoons by Jenny Gretton

HarperCollins*Academic*
An imprint of HarperCollins*Publishers*

Published by
HarperCollins_Academic_
77–85 Fulham Palace Road
Hammersmith
London W6 8JB
UK

First published in 1991

British Library Cataloguing in Publication Data

O'Connor, Maeve
Writing successfully in science.
I. Title
808

ISBN 0-04-445805-3
ISBN 0-04-445806-1 pbk

Library of Congress Cataloging in Publication Data

O'Connor, Maeve.
Writing successfully in science / Maeve O'Connor; with cartoons by Jenny Gretton;
sponsored by the European Association of Science Editors.
p.   cm.
Includes bibliographical references and index.
ISBN 0-04-445805-3 (acid-free). – ISBN (invalid) 0-04-45806-1 (acid-free)
1. Technical writing.   I. European Association of Science
Editors.   II. Title.
T11.033   1991
808'.0665–dc20

Typeset in 10 on 12 point Palatino by Computape (Pickering) Ltd, North Yorkshire
and printed in Great Britain by the University Press, Cambridge

Contents

Introduction

Are you a reluctant writer? If so, you are in good company. Many scientists, even the most successful ones, would rather get on with their next piece of work than settle down to reporting the last piece. But it is a fact of scientific life that every worthwhile research project must lead to publication or a written report of some kind. Each new step in science is based on earlier findings, and each step must be fully and clearly documented for the sake of others following the same path.

The first substantial piece of writing many scientists do is a progress report on their thesis work, or a short journal article written jointly with their supervisor, followed by the thesis itself. This book aims to make writing research papers, theses, and other kinds of reports easier, and perhaps more enjoyable, whether you are an experienced author or a beginner.

Research papers conventionally include an introduction, a description of what was studied and how it was studied, an account of the results, and a discussion of the results and their implications. Journal articles constructed in this formal way have become the basic units of research publication and are the model for many other kinds of writing in science. The chapters on writing journal articles therefore form the core of this book (in which 'science' means all branches of science, including medicine). Nearly everything in these chapters is relevant even if you are writing a thesis or a report rather than a journal article.

The first 11 chapters of the book constitute a complete revision of *Writing scientific papers in English* (O'Connor & Woodford 1975). These chapters cover all the steps in preparing a research paper for publication, from planning the paper to revising and typing it and checking the proofs. Chapter 12 discusses oral presentations and posters, because these play an important part in every scientist's career. The remaining three chapters deal with the particular requirements for writing theses, review articles and book reviews, and grant proposals and curricula vitae.

The book is for authors of any nationality but, like its predecessor, it pays attention to the needs of writers for whom English is not the first language. If you are such a writer, you should not imagine that your British and American colleagues receive preferential treatment from journal editors. Logical thinking and the logical organization of worthwhile subject matter are more important than an elegant literary style – and you may well have received a better grounding in grammar than some native English speakers. Scientists should aim for clarity, directness, and precision – that is, for good style but not necessarily for the heights of literary style. A well written article is bound to attract more readers than one that is badly written.

Before you start to draft a paper, read the most recent set of instructions or guidelines for authors prepared by the journal, institute, or agency for which you are writing. Those instructions or guidelines take precedence over any other advice, including the advice given in this book. The recommendations made here, however, are in line with recommendations published by the Council of Biology Editors (CBE Style Manual Committee 1983), the International Organization for Standardization (e.g. ISO 215: 1986, ISO 2384: 1977, ISO 5966: 1982, ISO 7144: 1986), and many scientific societies. Like its predecessor this book reflects editorial standards that have been adopted in most branches of science. I therefore hope it will smooth the rocky road to publication for you.

Acknowledgements

I am much indebted to Dr F. Peter Woodford, co-author of *Writing scientific papers in English*, whose plan for that book is observed in the first 11 chapters of this book. I am extremely grateful to those who read and criticized all or some of the drafts of the present book. Their comments and generous help were invaluable and I thank all the following reviewers very warmly indeed: Professor Knut Faegri, Dr Charles Hollingworth, Dr Joan Marsh, Mrs Julie Whelan, and Ms Mimi Zeiger. In addition I am grateful to Dr Douglas Altman and Dr David Strachan for their help with Chapter 5, to Dr Frans Meijman for suggesting and reviewing the section on review articles in Chapter 14, and to the participants in three courses on scientific writing organized by the European School of Oncology in Venice, 1988–1990, for their reactions to drafts of the book.

I am very grateful to Mrs Jenny Gretton, who drew the cartoons, and Mr Doig Simmonds, who supplied the other illustrations and much useful information for Chapters 4 and 12.

My thanks also go to the Council of the European Association of Science Editors (EASE) for supporting this successor to *Writing scientific papers in English* – which was itself initiated by EASE's parent, the European Life Science Editors association (ELSE), and sponsored by the Ciba Foundation, by ELSE, and by EASE's other parent, the European Association of Earth-Science Editors (Editerra).

CHAPTER ONE

Assessing your work and planning its publication

Dealing with preliminary questions   Writing and speaking while work is in progress   Assessing whether, when, and what to write   Deciding who and what to write for   Obtaining instructions to authors

Why must scientists write articles for publication? The reasons range from the pragmatic – such as winning fame, promotion, or a new job – to the idealistic – such as extending the boundaries of knowledge. The only good reason, though, is both practical and idealistic: communication is essential if science is to progress. You must document your work thoroughly before other scientists can repeat that piece of work, build on it, criticize it, or simply admire what you have achieved. In starting an investigation you are, in fact, taking the first step on the road to publication.

Research workers must write. They must also write simply and clearly enough for readers to understand and even enjoy what they are reading.

As a first step towards reaching the goal of comprehensible publication ask yourself the three sets of questions discussed below. The remaining steps in planning and preparing a research paper for submission to a journal are shown in Table 1.1 and described in Chapters 1 to 9. The steps in writing a thesis (Ch. 13) or technical report are similar to those in writing a journal article. You should of course adapt the sequence given here to your own way of working and to the kind of document you are writing.

THREE SETS OF PRELIMINARY QUESTIONS

First, are your studies designed to answer precisely the question you are examining? Have you drawn up a plan – a protocol – for what you intend to do? Do the studies cover all the criticisms likely to be made? Are the statistical methods valid?

1

Table 1.1 Main steps in preparing a research paper for publication

---

(1) Assess work (Ch. 1)
(2) Choose journal, obtain instructions to authors and read them (Ch. 1)
(3) Agree authorship (Ch. 2)
(4) Write working title and choose main headings (Ch. 2)
(5) Construct outline(s) (Ch. 2)
(6) Ask for permission to reproduce previously published material (if used) (Ch. 2)
(7) Choose and design tables and figures (Ch. 3 and 4)
(8) Write first draft and present preliminary paper (Ch. 5 and 12)
(9) Bury draft for a time, to give better perspective at revision stage (Ch. 5)
(10) Prepare reference list (Ch. 6)
(11) Revise structure (Ch. 7)
(12) Revise style (Ch. 8)
(13) Obtain comments from colleagues and revise again (Ch. 8)
(14) Reread instructions to authors, retype manuscript, and check final version (Ch. 9)

Secondly, do your experiments meet accepted ethical standards? You must answer this question if you are using human beings or animals, or if your work could affect the environment or the place where you are doing field work. You may also have to answer two related questions: is publication of your work likely to break any official secrecy regulations? and is publication likely to invalidate a later application for a patent? Take advice on these matters and, if necessary, modify your plans for what to include in the paper.

Thirdly, have you arranged how to record the details of your work as it proceeds? In some laboratories notes have to be kept in a prescribed manner, in bound notebooks with numbered and dated pages. Informal notes, including draft tables and figures, must also be kept in such a way that they can be found when wanted. From the beginning of an investigation you must therefore keep careful records, whether on paper or in a computer. Keep records of what you read (see Ch. 6), as well as of what you do.

WRITING AND SPEAKING WHILE WORK IS IN PROGRESS

When your investigation has been under way for some months, assess the direction the work is taking. One way to do this is to write down the question you are studying and describe what you have done so far. The act of writing about your work forces you to think about it more clearly than is possible while you are busy with the experiments or observations. Any lack of clarity shows up when you put your thoughts on paper:

. . . you must therefore keep careful records . . .

'writing necessarily uses words, and almost all thinking is done with words . . . the discipline of marshaling words into formal sentences, writing them down, and examining the written statement is bound to clarify thought' (Woodford 1967).

Another way to assess progress is to present your methods and results orally or in poster form at a departmental seminar, or at a meeting whose proceedings will not be published in full (see Ch. 12). The comments and questions at the meeting will help you to decide whether your work is ready for submission to a journal.

DECIDING WHETHER AND WHEN TO WRITE

After these preliminary assessments you must also decide whether your work is suitable for publication. Is it really worth writing about? Some investigations turn out to be unsuitable for publication even though the topic originally looked worthwhile. Aim to publish a few first-rate articles rather than numerous minor contributions; quality of content will do you more good in the long run than the number of papers to your name.

A paper worth publishing records 'significant experimental, theoretical or observational extensions of knowledge, or advances in the practical application of known principles' (O'Connor & Woodford 1975, p. 3). Do your results and conclusions fulfil one of those requirements? One way of deciding whether they do is to write down the conclusions you have reached. Measure these tentative conclusions against what is known about the topic and then, if possible, show them to an experienced colleague working in a different area of science. Discussing your conclusions with an unbiased person, like making a preliminary presentation at a meeting, will give you a fresh view of your findings and how to interpret them. Putting your conclusions on paper and discussing them will also reveal whether you need to do more work to fill gaps in your arguments or observations before starting to write.

Another question to ask yourself is whether your tentative conclusions are still clearly related to the question you set out to study. If your findings have led you away from the original question, rethink and restate the question now, to prevent confusion in what you write later on.

After you have answered these questions satisfactorily, the point at which you start to draft the paper will probably be influenced by down-to-earth events. You or a colleague may be about to leave for a new job, grant money may be running out, or a deadline may have to be met. Whatever the circumstances, make an early start on sections such as the methods or results, but don't begin the major job of drafting the whole paper until you are sure your findings are reasonably firm and complete, take knowledge of the subject a step further, and are 'new, true and meaningful' (DeBakey 1976).

DECIDING WHAT KIND OF PAPER TO SUBMIT FOR PUBLICATION

You also need to decide whether your material is suitable for a long paper with a detailed discussion or for a short report of results, with brief comments, if any. Would it be better to send a full report to a conventional journal, or submit a brief note or preliminary paper to a journal specializing in rapid communication, if that is customary in your discipline? If the paper is the first on which you are the first or sole author, aim to make it as complete and as good as possible. If you are working in a very competitive field, however, consider whether to write a shorter paper for a 'letters' journal.

Don't slice a single piece of work into several short papers ('minimum publishable units') without a good reason for doing so. Editors call this kind of serial submission 'salami science' and they tend to reject the later slices. Publication of several papers on the same piece of work is acceptable only for large-scale investigations taking several years to complete, or for investigations that fall naturally into several self-contained parts or into parts that interest readers in different disciplines.

A paper you submit to a journal must be based on your own original work. Passing off another person's work as your own – plagiarism – is fraudulent and unacceptable. The paper must also be new: that is – with

4

the exceptions discussed below – it must not have been published in or concurrently submitted to another journal. Publishing similar papers in two or more journals – duplicate or multiple publication – is nearly always as unacceptable as plagiarism. Concurrent submission is unacceptable too, with the exceptions discussed below. Submitting the same paper to more than one journal at once may seem a good way of saving your time but it wastes time for editors and referees, and money for journal publishers. Many journals therefore ask authors to state in a covering letter or on a copyright form that the paper has not been published elsewhere and has not been concurrently submitted to another journal.

The rule about multiple publication does not apply to preliminary communications. If you decide to write a brief note or preliminary communication you can still submit a full paper on the same topic later, provided that the preliminary version is not too detailed and that you refer to its existence in the later, fuller, version. Abstracts of presentations at scientific meetings don't count as 'previous publications' either, provided that they are not too detailed and that not too many details have been released to the media, for example at a press conference, and disseminated widely before the full paper can be published.

Other exceptions to the rule about duplicate submission or publication may arise when the journals are aimed at totally different audiences, such as clinicians and geographers, or are in different languages, or – if both journals are local or regional rather than international – are intended for distribution in different geographical areas. If you think your paper will be suitable for duplicate publication under one of these three conditions, write to the editors of both your target journals at an early stage and obtain their consent to duplicate submission.

DECIDING WHO THE READERS WILL BE AND CHOOSING A JOURNAL

All writers, including scientists, must keep their readers in mind while they are writing, so consider next who your readers will be and why they will want to read your paper. Thinking about your audience will help you to choose the journal to which you will submit the paper. Choosing the journal will also enable you to look at examples on which you can model the paper and obtain instructions on how to prepare it.

Don't necessarily choose the first journal that comes to mind – usually the one you read most often. Ask yourself whether the work you are reporting will interest a wide general audience, or everyone working in your branch of science, or just a few specialists. Then draw up a short list of possible journals, based on those considerations and on other points such as the following:

(1) *Which journals do the people you hope to reach prefer to read?*

Ask your more experienced colleagues about this point, or consult an information scientist in your institution's library. If your first language is not English, consider the language of publication too: your paper may remain hidden for years if you publish in a minority language. (But balance this against point 4 below.)

(2) *Which journals print papers of the kind and length you propose to write?*

The scope of a journal is usually defined at the beginning of the instructions to authors but is sometimes described in or near the masthead of the journal (the section where the publisher, editorial staff, frequency of publication, and subscription rates are listed).

(3) *Which are the well-established journals in your field?*

Be careful about submitting papers to new journals or journals still in the gestation period. If such a journal fails, your paper may die with it. On the other hand, if the subject area covered by the journal is exactly right and the editor and publishers seem sound (ask colleagues for their advice), a new journal may be a good choice. Be careful, too, about submitting your early papers to 'throwaway' journals. Publishing in these is unlikely to help your career.

(4) *Which journals are of high scientific quality but have moderately low rejection rates (say 30% of submissions, or less)?*

Rejection rates vary widely both between and within disciplines. International journals of high prestige may have rejection rates as high as 80–90% of submissions. You might be better off choosing a national or local journal with a lower rejection rate, especially if your paper is of national or local interest. Some journals publish their rejection rates annually, or more experienced colleagues may be able to tell you what the rate is for a particular journal.

(5) *Which journals are covered by the main abstracting and title-listing services?*

Journals may name these services in or near the masthead, or in publicity material. Or see the most recent edition of *Ulrich's international periodicals directory.*

(6) *Which journals have editors who are highly regarded in their fields of science and provide prompt, fair, and helpful reviewing?*

Not all first-rate research workers make first-rate editors, but most work hard to keep their reputations by producing the best journals they can. A star-spangled editorial board or advisory committee does not, however, guarantee that the editorial system is first-rate or that the referees are well chosen. Ask experienced colleagues for their views.

(7) *Which journals are published often enough to give your paper a chance of appearing within six weeks to nine months of acceptance (depending on whether the journal is a rapid results journal or one with a slower publication schedule)?*

A monthly journal may be a better choice than a quarterly, all else being equal, but if the journal publishes acceptance dates it is worth calculating the publication lag – the time from acceptance date to publication date. You can minimize delays by following the journal's instructions closely (see 'Obtaining and reading the instructions to authors' below).

(8) *Which journals require authors to be members of the society which owns or sponsors the journal?*

The instructions to authors will answer this question.

(9) *Which journals have page charges or submission charges?*

If you can't find funding for page or submission charges this point could be important. Some journals waive these charges in certain circumstances, for instance for authors from countries with currency exchange problems. If your paper is well suited to such a journal, consult the editor about the charges if payment is a problem.

(10) *Which journals print high quality photographs (e.g. electron micrographs), and which accept colour photographs, if these are essential to your work?*

Your own and your colleagues' observations will tell you which journals are suitable.

(11) *Which journals provide offprints or reprints, which provide them free of charge, how many do they provide, and what do extra copies cost (if offprints/reprints are important to you)?*

You may wish to publish in a less well-known journal with a generous reprint policy that enables you to reach exactly the readers you want by sending them offprints or reprints. (See p. 148 for the distinction between offprints and reprints.)

(12) *Which journals use a standard reference system that you can cope with easily?*

This seemingly minor point will become a major annoyance if you haven't got a bibliographic program for formating references on a computer and if a lot of trivial changes have to be made which don't help readers at all – such as changing 1990a, b, etc. to 1990*a*, *b*, etc. throughout, or changing Jones B, Smith A, etc. to Jones B, A Smith, etc.

OBTAINING AND READING THE INSTRUCTIONS
TO AUTHORS

When you have chosen a journal from your short list the next step is to read the current version of its instructions to authors. Many journals print these instructions (or guidelines, or information) in every issue, some print them once a year in the first issue of the year, and a few print them as a separate leaflet or booklet which you can obtain, usually free, by writing to the editorial office. If you work in a biomedical discipline you will find that many journals use, or say they use, the 'Uniform requirements for manuscripts submitted to biomedical journals' (Appendix 1) (ICMJE 1988), a set of instructions intended to allow authors to use the same format and style for papers submitted to different journals.

Editors enforce instructions because the instructions are designed to promote swift and accurate publication and to save money. For instance, the common requirements for double spacing and margins of at least 25 mm reduce the chances of embarrassing errors slipping into print and reduce editing and typesetting costs. The time you spend getting the format and style right will be trivial compared with the time and money spent on the research you are describing. The better you observe the instructions the more likely the journal is to make a good job of producing your paper.

The journal's instructions will probably state a minimum and maximum acceptable length for papers and give other information about physical format, such as whether papers should be in double spacing on A4 paper, or whether submission on a floppy disk or magnetic tape, with or without a printout, is acceptable or essential. (See Ch. 9 for more about the physical format and technical preparation of papers.)

The instructions to authors usually specify how tables, figures, and references should be presented. They may remind you to use the international system for units of measure (Système International d'Unités: SI), and they may state which rules of nomenclature to use, or which style guides to follow. You should obtain and read any documents that are recommended.

The instructions may include other requirements that are not necessarily relevant at this early stage of preparing for writing but which you should note for later action. For example, you may be asked to submit copies of permission letters from ethics committees if your paper describes experiments with human beings or other animals. You will probably have to provide copies of letters giving permission to reproduce material from other people's work (see Ch. 2, 'Borrowing published work'). You will probably also be asked to assign your copyright in the paper to the publisher by signing a copyright form, either when you submit the paper or when it is accepted for publication (see p. 18 and p. 131).

THE NEXT STAGE

When the preliminary steps described in this chapter are complete, you will be ready to begin constructing the framework of the paper, consisting of headings and outlines (p. 12–15) and tables and figures (Chapters 3 and 4).

SUMMARY

(1) Answer preliminary questions about the direction, content, and organization of your work; (2) Assess whether your work is ripe for submission; (3) Decide whether to write a short paper or a more detailed one; (4) Decide who your readers will be and which journal you will submit the paper to; (5) Obtain the journal's instructions to authors and read them carefully.

CHAPTER TWO

Getting started: building the framework

Agreeing on authorship   Writing a working title   Deciding the structure of the paper (choosing its main headings)   Constructing outlines   Obtaining permission to borrow other people's work

When you have assessed your material and chosen a target journal you will be ready to design an effective paper. Begin by agreeing on its authorship and giving it a strong framework.

AGREEING ON AUTHORSHIP

If you have co-authors, problems about authorship can range from the trivial to the catastrophic. You should therefore discuss the question of authorship before the drafting stage. Anyone named as an author should have made a substantial contribution to the work being described. In the words of the 'Uniform requirements for manuscripts submitted to biomedical journals' (Appendix 1) (ICMJE 1988):

> Each author should have participated sufficiently in the work to take public responsibility for the content.
> Authorship credit should be based only on substantial contributions to (a) conception and design, or analysis and interpretation of data; and to (b) drafting the article or revising it critically for important intellectual content; and on (c) final approval of the version to be published. Conditions (a), (b), and (c) must all be met.

The 'Uniform requirements' also state who should *not* be included as an author:

> Participation solely in the acquisition of funding or the collection of data does not justify authorship. General supervision of the research group is also not sufficient for authorship.

That is, you should not include the head of your department or the Nobel Prize winner in the next laboratory or any other senior colleague as an author unless that person chose the topic of your research, or planned the experimental approach, or made some other substantial intellectual contribution to the work – but don't lose your job arguing this point. Don't include as authors people who supplied you with material or simply advised you while you were doing the work, and don't include technical or other staff who helped you as part of their normal duties. Instead, use the acknowledgements section of the paper to thank anyone whose help you want to recognize (and obtain their approval of the way you thank them – see Ch. 5, 'Acknowledgements').

Everyone listed as an author must agree to be named as an author – never include people without their knowledge.

Journal editors tend to become suspicious if the number of authors is more than is normal for the discipline. Some journals, particularly those that follow the 'Uniform requirements' recommendations carefully, ask all authors to affirm that they contributed substantially to the work and can take responsibility for it if their co-authors are not available to do so. Some journals also ask authors to justify their authorship by stating the part each contributed to the paper.

Agreeing the order of authors' names

If you work in a large team and are not one of the first three authors named, be prepared for your name to be invisible in reference lists in other authors' articles. Some journals print all the names in reference lists if there are up to perhaps six authors but print only the first three names if there are seven or more authors. Even when the name-and-year system (see Ch. 6) is used for references you are unlikely to be named in the text if you are not the first author, because it is quite common for only one author to be named when a multi-author paper is cited. It is therefore important to agree not only who the authors are, but also the order in which their names will appear on the title page of the manuscript. Some journals ask for the names to be given in alphabetical order – but ignore such a request unless all the authors agree to it and all of you contributed equally to the paper. The authors themselves should usually decide the order.

If one author does most of the work and all or most of the writing, that person's name should go first. The other names should be listed in an agreed order reflecting how much each person has contributed to the work. A common convention is for the leader of a research team or the head of the department to go either last or second in the list of authors – provided, of course, that he or she has contributed significantly to the

11

work. If everyone contributes equally to the paper, arrange the names alphabetically, or follow local or national conventions.

Another convention is for members of teams who write several papers together to take it in turn to be named as the first author. But avoid naming different first authors for papers that have the same general title with numbered subtitles: series of this kind with different first authors are a librarian's nightmare and are unpopular for other reasons too (see Ch. 5, 'Title').

You should also find out exactly how your co-authors like their names to be written (see Ch. 9, 'Title page').

Writing as one of a team

If you are working with several co-authors, one person in the team should, ideally, be responsible for doing the writing. If it is more practicable for different people to write different sections, one person should edit the final draft to give the paper the necessary unity and consistency.

Team writing requires careful planning. Woolston et al. (1988) recommend that the group should agree on the purpose, scope, and outline of the paper before drafting starts. These authors suggest using a 'writing unit' of perhaps two to five pages, with everyone assigned to write the same number of units so that everyone writes to the same level of detail. They also point out that meetings to discuss the draft are essential but should be kept short and efficient. Each person should have sole responsibility for his or her section(s) of the work, with progress being reviewed frequently in the development stages.

All the authors in a team should read and approve the final version of the manuscript before it is submitted for publication. They must also agree, later, on changes to be made in response to referees' comments.

WRITING A WORKING TITLE

Writing a working title before you draft the rest of the paper will help you to define the scope of the paper. Write a title that describes your main subject, not the minor topics that may also be covered. The title can be any length you like at this stage – you will want to rewrite it later (see Ch. 5, 'Title').

DECIDING THE STRUCTURE: CHOOSING THE MAIN HEADINGS

Headings in a paper are reference points that make the structure of the paper clearer to the writer as well as to the readers. In practice you may have chosen some headings when you were recording your work and making early versions of tables and figures. Now write or revise them with publication in mind, wording them informatively.

Your most likely choice for the main headings, especially in biomedicine, will be the conventional Introduction, Methods (or Materials and methods), Results, and Discussion – often called the 'IMRAD' structure. In a descriptive field science the main headings might be Introduction, Geographical or historical context, Field work, Analysis of results, Discussion, and Conclusions. For a theoretical paper the main headings might be Introduction, Theoretical analysis, Applications, Conclusions. A paper describing a new method might have the headings Introduction, Description of procedure, Tests of new method, Discussion.

Results are sometimes combined with the discussion section but this can confuse readers and it is better to keep these sections separate, especially in your first papers. In some branches of science it may be more effective to present the results immediately after the introduction, to explain why you chose the methods you go on to describe. Less often, it may be useful to combine methods and results.

If articles in your target journal usually include a section headed conclusions, or conclusions and summary, plan the end of the discussion section appropriately.

Rather than putting detailed descriptions of procedures or other lengthy details in the body of the paper you might want to put them in an appendix. If so, this must be part of your plan for the paper, not an afterthought submitted with the proofs, because an appendix has to be reviewed with the rest of the paper. Alternatively, you may be able to send lengthy material to an archive recommended by the journal. Other possibilities, such as including methods in table footnotes, figure legends, or separate sections at the end of a paper, depend on journal practice.

Articles in recent issues of your target journal, with the instructions to authors, will show you what kind of structure is most commonly used in the journal. If your material does not fit happily into the conventional scheme, choose more appropriate headings – few journals have strict rules about what the headings should be. But don't stray too far from the usual pattern for your discipline without having good reasons for doing so.

13

When you have chosen the main headings a good way to proceed is to write each heading at the top of a sheet of paper (Woodford 1968). Then list all the points you can think of that seem to belong under each heading. Write the points down in whatever order they come to mind, without trying to arrange them in any logical way. Next find all your notes, quotations, reference details, and any figures and tables you have already drafted. Sort these out and put them with the relevant sheet of paper in your filing system, or, if everything is in your computer, move the various parts to the appropriate heading or file. You will then be ready to make an outline of the paper.

CONSTRUCTING OUTLINES

An outline organizes your argument and makes for a tighter, more comprehensible, paper. It provides a route map that will get you back on course easily after interruptions or delays in the drafting process. Constructing an outline is therefore an essential step in preparing work for publication.

An outline can be either a topic outline or a sentence outline. A topic outline is a logically ordered list of the points to be included in a paper; it consists of a series of nouns or phrases that indicate what you will be discussing. A sentence outline expands those nouns or phrases into sentences that say what each paragraph discusses; that is, the sentences will later become key or topic sentences for paragraphs in your draft. A topic outline on its own may be enough, or you may prefer a mixed topic and sentence outline.

Topic outline

Construct a topic outline by sorting the points listed under your main headings into a logical order. Look through your notes and other material to see if any extra points should be added now. Then assign the topics to different levels according to their importance, preferably using not more than two or three levels under each main heading. To make it clear which level of heading is which, use either indentation or a numbering system (but not necessarily both, though both are shown in the example below). An ideas organizer/outlining program may be useful at this stage, if you have one for your computer.

A topic outline for a short article on the preservation of plant germ-plasm, for example, might look like this after the points have been ranked in logical order under the main headings:

1.0 Introduction
   1.1 Traditional methods
   1.2 New methods
   1.3 Plant tissue culture
2.0 Methods and Results
   2.1 Minimal growth conditions
      2.1.1 Systematic study lacking
   2.2 Undercooled conditions
      2.2.1 Fragmentary information
   2.3 Low temperatures
      2.3.1 Organ cultures
      2.3.2 Callus cultures
      2.3.3 Liquid nitrogen temperatures
      2.3.4 Ice crystals/dehydration
      2.3.5 Cryoprotectants
      2.3.6 Other aspects?
3.0 Discussion/Conclusions
   3.1 Vials in refrigerator
   3.2 Cryogenic storage possible
   3.3 Funding for evaluation

Sentence outline

A sentence outline based on the topic outline above might read as follows:

Traditionally, plant germplasm is stored in seed collections, arboreta, and so on. New methods of storage now being explored are based on plant tissue culture. Maintenance of tissue cultures under minimal growth conditions appears promising but has had little systematic study. Undercooling plant cells and tissues to $-40\,°C$ also holds possibilities but information on this method too is fragmentary. Storage in liquid nitrogen has been successful with organ cultures from 10 species. Callus cultures from a dozen other species have also been successful. Temperatures of $-196\,°C$ or $-140\,°C$ seem to be needed for successful storage, to prevent cells from being destroyed by the formation of intracellular ice crystals or by excessive dehydration. Cryoprotectants such as DMSO or proline are needed to enable cells and tissues to survive freezing and thawing. Other important aspects affect survival. Large numbers of genotypes could be kept in ampoules maintained in liquid nitrogen at $-196\,°C$ or $-140\,°C$ in a small refrigerator. Cryogenic storage of plant germplasm is feasible but the technique needs further exploration and development. Funding to evaluate the technique will depend on the demand for such a method of storage.

COPING WITH COPYRIGHT

By the time the outline is complete you may know whether you want to borrow material from published or unpublished work by other people. Written permission has to be obtained to reproduce tables and figures and to quote someone else's words directly – unless the material is in the public domain in the United States, when you may use it freely. It is your responsibility, not that of the editor of your target journal, to obtain a 'release' letter from the copyrightholder. File such letters when you receive them and make copies when you need them.

Borrowing published work

If no waiver or word limit is mentioned in the original publication, ask permission from the copyrightholder when you want to reproduce a complete table or figure or quote more than, say, 100 words or 5% of the original publication, whichever is less. This permission is usually given freely and without charge. If the copyrightholder makes a charge you may decide to redraw the original material or rewrite quotations in your own words rather than pay to use the original. If you redraw or rewrite material you must still include a reference to the original work. If you redraw a figure, for example, write 'Based on [or Redrawn or Adapted from] Brown (1988)' in the legend and include the reference in your reference list.

The copyrightholder of a journal article is more often the publisher than the author but may sometimes be another person or organization named on the title page of the article or on one of the early pages of the original publication. If it is not clear who the copyrightholder is, you can nevertheless write to the Permissions Department of the publisher or to the editor of the journal.

If the author of the material you want to borrow is not the copyright-

---

Figure 2.1 Sample letter to copyrightholder and author, asking for permission to reproduce material from a journal article in the article you are writing. If you want to use the borrowed material in a book, include the name of the publisher, the place and expected date of publication, and preferably also the number of copies to be printed and the type of rights requested – for example, non-exclusive English-language rights world-wide. (Publishers usually supply authors with a copy of a suitable permissions letter giving the necessary information.)

Dr A. Smith
Editor, Space Exploration
1 The Buildings
University of Loamshire
Loamtown
LO1 3UP

Dear Dr Smith

I am preparing an article [tentatively] entitled

Travels in outer space

for submission to:

Out of Bounds

I should be grateful for your permission to include the following material:

Figs 1 and 2 from your paper/the paper by J. Walker entitled 'Going far?', published in Space Exploration 1987; 16:150–165.

I am also writing to the copyrightholder/author requesting permission to reproduce this material. A full reference to the original paper will be included.

The acknowledgement will include the words 'Reproduced by permission of [copyrightholder]' and I should therefore be glad if you would confirm the name of the copyrightholder. If this form of acknowledgement is not sufficient, please indicate below what form the credit line should take.

Please indicate your agreement to this request by signing and returning the attached copy of this letter.

With many thanks for your help,

Yours sincerely

W. Brown

I/We give permission for this material to be used as specified above:

Signed: .... (Copyright-holder/Author)

Name: ....

Date: ....

Credit line to be used:

....

---

holder it is polite to obtain his or her approval as well as that of the copyrightholder. One reason for writing for permission at an early stage is that authors sometimes write back with new information that may be important for your work. Another reason is that copyrightholders often take three to six months or more to reply, though permission is seldom refused.

You stand a better chance of receiving a prompt reply if you send the copyrightholder and the original author two copies each of a letter/ release form such as the one shown in Figure 2.1, together with a self-addressed envelope, stamped if possible. Enclose a photocopy of the material you want to borrow, showing any minor changes you plan to make. If the wording of the acknowledgement or credit line is not specified on the release form when it is returned to you, write 'Reproduced, with permission, from [Author name; Year; Journal name; Vol. no; Inclusive page nos.]' after the quotation or in a table footnote or figure legend. Include the full reference in your list of references.

There is no need to obtain written permission when you are simply referring to other people's work rather than borrowing pieces of it. But you must of course name the source of work you refer to, including the source of any parts you rewrite in your own words or include in tables or figures. Cite the original author(s) appropriately in the text, or in table footnotes or legends for figures, and include details of the original work in the reference list.

If you want to reproduce material from an earlier article of your own for which you signed a copyright form (see below), check the wording on a copy of the form. Some publishers allow authors to re-use their own work without asking permission. If no such waiver appears on the copyright form you must obtain written permission to re-use your own material too.

Borrowing unpublished work

Permission is also needed to refer to or reproduce unpublished work. The original authors or speakers may not want you to mention what they said in letters or conversations, or in lectures or informal discussions at meetings that are not being published. Citing unpublished work is discussed in Chapter 6.

Signing copyright forms

The other aspect of copyright, assigning rights in your article, need not concern you until you submit the manuscript or until the manuscript has been accepted for publication. At one of those stages you will usually be asked to sign a form transferring copyright to the publisher or giving the publisher a licence to publish the work. A transfer or licence allows the publisher of a 'collective work' – a journal or a multi-author book – to reply to requests for reprints or for permission to reproduce parts of the work, as well as to handle other questions affecting rights in the work. In general, transferring copyright to the publisher does not harm the interests of authors of contributions to journals or multi-author books. If you are writing a book on your own or with only one or two other authors, however, get advice from a lawyer or from experienced authors about the contract the publisher offers you. If you are writing a potential bestseller, find an agent to advise you about publishers and contracts.

TAKING THE NEXT STEPS

Before drafting the text of your paper you should draft the tables and choose the figures needed for it (see Ch. 3 & 4). Tables and figures, which contain the actual results, are the heart of most research papers and you must make the text fit the results, not the other way round. Being able to look at the draft tables and figures when you are writing the text will also help you to maintain consistency between text, tables, and figures.

SUMMARY

(1) Agree who the authors are; (2) Decide the order of authors' names; (3) Agree how to work with your co-authors; (4) Write a working title; (5) Decide the structure of the paper by choosing the main headings; (6) List the points that belong under each heading; (7) Construct a topic or sentence outline of the paper; (8) Write for permission to borrow other people's work, if you want to include such work in your paper.

CHAPTER THREE

Preparing effective tables

Designing preliminary versions   Size and format   Title   Column
headings   Field contents   Explanatory notes   Making final
versions

Readers often look at tables and figures to see whether the rest of a paper
is worth reading. Each table should therefore be capable of standing on
its own, without reference to the text. Each table must also earn its place
by contributing an essential part of the story you tell in the text.

Decide first whether to present your findings in a table or a graph.
Tables and graphs cost a similar amount to produce but both cost more
than text; the editor may remove some if you exceed the journal's limit.
Tables are better for reporting precise numerical information which can
be directly compared with numbers reported in other publications, for
recording numbers when there are not enough to make a satisfactory
graph, and for showing data for component groups. Graphs are better for
illustrating trends and the relations between variables in experimental
data, and for recording numerical findings when there are a lot of these
(CBE Style Manual Committee 1983, p. 67).

DESIGNING PRELIMINARY VERSIONS

Check the instructions to authors to see whether the journal limits the
number or size of tables.

You will probably make rough versions of tables as your work
proceeds. The best way of recording work in progress, however, may not
be the best way of presenting the same work in print. Before drafting the
text, decide which tables you need for the paper and redesign them with
publication in mind, as described in this chapter.

Design separate tables for separate topics. Don't use tables just to show
off how much data you have collected; instead, give sample data. Don't
repeat data in tables if you plan to put the same data in the text or in a
figure.

Size and format

A table consists of a title, column headings, row (or side) headings, the field (the rows and columns containing the data), and, usually, explanatory notes. With a framework of this weight a table must contain enough data to justify its existence. If your proposed table has only one or two rows of data, present your findings in one or two sentences in the text instead of constructing a table – unless those findings are so important that readers ought to see them when they first glance through the paper. On the other hand, don't make tables so large that the journal page cannot hold them. If you are listing features in words rather than numbers, consider whether you really need a table; a few sentences in the text may be better. For example, the table in Figure 3.1 could easily be converted to text in the following way, unless you have a compelling reason for putting the information in a table:

Species A in Lake 1 was affected by diseases X and Y, species B in Lake 2 by disease Y, and species C in Lake 3 by diseases X and Z.

Table 1   Diseases in three species of fish in lakes in North Wales.

Disease	Species	Habitat
X	A	Lake 1
X	C	Lake 3
Y	A	Lake 1
Y	B	Lake 2
Z	C	Lake 3

Figure 3.1   A table that would be better converted to text.

Many tables in scientific papers follow the pattern shown in Figure 3.2, with three full-width lines ('rules') and shorter rules ('straddle' or 'spanner' rules) to clarify which columns the headings apply to. Use tables published in your target journal as models for the draft tables.

When you design tables keep the structure as simple as possible. Ask yourself what message you want readers to extract from each table and what information those readers already possess. Arrange the data accordingly and in a way that matches the way most people read tables – usually from left to right and from top to bottom.

If readers are likely to look for a particular item, for example a protein, and then read off values for a certain characteristic of that item, such as its

Table 2 Mean percentage of true digestibility and net utilization of protein in seed meals.

Stub or side heading	Column heading (straddle)		Column heading (straddle)	
	Column subheading (units)[a]	Column subheading (units)	Column subheading (units)[b]	Column subheading (units)
Row (stub) entry 1, indented for run-over	98.19* (0.34)	... [c]	70.99* (0.41)	86.84* (0.13)
Row 2 entry	77.23** (0.56)	96.91* (2.17)	65.76** (0.53)	91.32* (0.44)
Row 3 entry	89.45* (0.30)	78.87* (1.33)	81.23* (0.44)	79.82* (0.35)

The χ^2 test was used to measure the difference between observed values and expectations.

Standard errors of the mean are given in parentheses.

[a] Footnote a.

[b] Footnote b.

[c] No measurements made.

* $P < 0.05$, ** $P < 0.01$.

Figure 3.2 Typical table with three full-width rules and straddle (or spanner) headings. Probability values are expressed conventionally, but Gardner & Altman (1989) recommend that exact values be given. The asterisk (star) convention saves space and can be read at a glance, but follow your target journal's practice in statistical matters.

concentration, list all the items vertically to form the first column (the 'stub'; the entries in the stub are the row entries or row (side) headings, as shown in Fig. 3.2). The names of the characteristics (concentration, molecular mass, etc.) then head the remaining columns. If you think readers are more likely to look up a characteristic such as molecular mass and then read off the value for each of several items, move the characteristics to the stub. Nevertheless, if there are more items than characteristics, list the items in the stub, to save space (Reynolds & Simmonds 1981).

Try to design tables that fit easily into one or more journal columns. If the journal has single-column pages don't make narrow tables unless they are so narrow that two will fit side by side on the page. (You can estimate the width of the tables by counting the number of characters – letters, numbers, symbols, and spaces – in the longest line of each table, allowing two spaces between columns. Compare the resulting character count with the number of characters in tables in the journal.)

If a draft table seems too wide, ask yourself whether all the columns are needed. For example, if all the entries in a column are the same (Fig. 3.3), put the information in the explanatory notes or in the table title (e.g. the sex of an experimental animal) (Fig. 3.4). Similarly, transfer a column of reference values or a column headed 'Notes' or 'References' to the explanatory notes. If column 5 contains values that can easily be calculated or deduced from the values in columns 3 and 4, omit column 5. If two or more row headings are repeated several times, change them to centred subheadings (Fig. 3.4) in the body of the table.

If these methods of reducing the width don't work, try splitting a very wide table into two smaller tables (but not too small). Alternatively, send a large collection of data likely to interest only a few readers to an archive for storage separately from the published paper (the journal's instructions to authors will tell you whether such storage is available). Make sure, however, that the tables you submit for publication contain enough values for referees to assess what you have done and for other workers to check their results against yours.

As well as making tables of a suitable width, make them a suitable length – neither too long nor too short. Editors may ask you to shorten tables that would stretch over two or more journal pages; count the

Table 3 Rates of acceptance of manuscripts by authors from different regions and disciplines after manuscripts had been incubated for different periods (inc. period) before submission to journals.

Region	Discipline	Sex of authors	No. of MSS/inc. period	Length of MSS (av.)	Length of incubation				
					0 days	1 day	1 wk	2 wk	3 mo.[a]
North America	Biochemistry	M	300	20 pp.	1		3	30	50
,,	Botany	M	250	22 pp.	1		2	24	48
,,	Chemistry	M	200	10pp	0		3	25	46
,,	Sociology	M	900	99 pp.	0		0	10	20
,,	Zoology	M	500	25 pp.	2		4	29	52
Europe	Chemistry	M	500	22 pp.	4		6	34	52
,,	Biochemistry	M	300	15 pp.	3		4	35	54
,,	Sociology	M	800	90 pp.	0	0.5	11	25	34
,,	Botany	M[b]	250	20 pp.	1		3	28	53
,,	Zoology	M	500	23 pp.	3		6	32	56
[etc.]									

[a] The lack of further improvement after incubation for three months is attributed to loss of an unknown number of manuscripts or loss of interest by their authors.
[b] Ten of these authors were women.

Figure 3.3 Table with poor title, too many columns, unnecessary repetition, irrational and inconsistent order for row entries, and no indication of what the numbers in the last five columns mean. See Figure 3.4 for improved version.

Table 4 Effect of incubation on acceptance rate of manuscripts from male authors[a] from different regions and disciplines.

Discipline & no. of MSS	Mean no. of pages	% Acceptance after incubation				
		0	1	7	14	92[b] (days)
		North America				
Botany (500)	25	2	4	29	52	49
Zoology (300)	20	1	3	30	50	45
Biochemistry (250)	22	1	2	24	48	42
Chemistry (200)	10	0	3	25	46	43
Sociology (900)	99	0	0	10	20	35
		Europe				
Botany (500)[a]	23	3	6	32	56	50
Zoology (300)	15	3	4	35	54	52
Biochemistry (250)	20	1	3	28	53	49
Chemistry (500)	22	4	6	34	52	47
Sociology (800)	90	0	0.5	11	25	34

[a] Ten authors of botany papers from Europe were women.
[b] The lack of further improvement after incubation for three months is attributed to loss of an unknown number of manuscripts or loss of interest by their authors.

Figure 3.4 Improved version of Figure 3.3: two headings have been moved to the body of the table, two columns have been amalgamated, a column whose entries were all the same has been removed, 'pp.' has been moved to the column heading, the straddle heading explains the entries in the last five columns, and the entry with a decimal point is properly ranged. Note that this table illustrates format only; an editor would reject it for its content.

number of rows in a large table in your target journal to see what the maximum length can be.

Don't try to cram a table onto the page by using narrow spacing, tiny print, or photocopier reduction. The copy editor needs room to insert typographical instructions, and small print is more likely to attract typesetting errors. If you submit complicated tables you may be asked to prepare them in such a way that the manuscript versions can be used as 'camera-ready copy' (material photographed directly for printing purposes rather than being typeset first). When you prepare such a table, make the lettering suitable for reduction to 50–75% of the original size (see Ch. 4, 'Lettering and symbols').

Table titles

Draft a concise title for each table. State the point of the table or say which items it compares, and perhaps indicate the experimental design. Say what the table shows, not simply what it's about ('Increasing incidence of disease in three species of fish in lakes in North Wales, 1980–1989', not 'Diseases in fish in Wales'). Aim for consistency in style and length of titles, bearing in mind the type and length of table titles in your chosen journal. Make a list of the titles to help you to see which tables are essential and whether their collective message will be clear to readers. Note that some journals allow or require brief descriptions of the methods to follow the title, as mentioned under 'Footnotes/headnotes' below.

Column order and column headings

Arrange the columns in each table in the order that should be easiest to understand and that shows readers what conclusion(s) you want them to draw. Put columns that need to be compared next to each other. Remember that it is easier to compare a series of numbers going down a column than a series of numbers going across a row (CBE Style Manual Committee 1983, p. 76–77) (Fig. 3.5). Put control or normal values near the beginning – usually in the first row or as the first column of data. If it seems logical to arrange the numbers from small to large, do so, or find some other logical way to group the findings. It is neither logical nor helpful to readers if you arrange entries in the order you did the experiments, or according to some personal numbering system, or by alphabetical order.

Make the column headings as brief as possible, to save space. The row headings in the stub can be longer than the column headings and can run over more than one line (as in Fig. 3.2), but they should nevertheless be kept short too. If the same words appear in column headings and in the row headings, see whether some of the words in the column headings can be left out.

Avoid using numbers with multipliers (exponents) in column headings: in a column headed '$\times 10^3$' the entry '5.0' could mean either 5000 g or 0.005 g. Make your meaning clear by putting 'kg', 'mg' or whatever unit is appropriate instead of '$\times 10^3$' at the top of the column.

Contents of the field

The field is the part of a table that contains the data, arranged in columns and rows (Huth 1987). Give numbers to an appropriate degree of accuracy – one that does not imply greater precision than you achieved with your measurements. In biological work, for example, the use of four or more digits is rarely justifiable.

Give numbers to the nearest significant figure, rounding the last digit up or down as necessary: when the last digit is 5 round the figure off in the direction of the even number (1.565 becomes 1.56, 1.575 becomes

Table 5a Decrease in twitch rate in patients with restless finger syndrome as concentration of Substance Z increases.

	Concn of Substance Z (M)		
	10^{-6}	10^{-5}	10^{-4}
No. of twitches/120 s			
Control	43 ± 4	40 ± 7	41 ± 3
After 60 s of drug	42 ± 5	30 ± 3	10 ± 2
Displacement of forefinger (cm)			
Control	0.5 ± 0.09	0.4 ± 0.07	0.4 ± 0.09
After 60 s of drug	0.5 ± 0.02	0.2 ± 0.04	0.1 ± 0.01

Three groups of 10 patients (5 F, 5 M; age range 59–73 years) were tested. Each group acted as its own control. Values are means ± standard errors.

Table 5b Decrease in twitch rate in patients with restless finger syndrome as concentration of Substance Z increase.

Concn of Z (M)	No. of twitches/120 s		Displacement of forefinger (mm)	
	Control	After 60 s of drug	Control	After 60 s of drug
10^{-6}	43 ± 4	42 ± 5	5.0 ± 0.9	5.0 ± 0.1
10^{-5}	40 ± 7	30 ± 3	4.0 ± 0.7	2.0 ± 0.4
10^{-4}	41 ± 3	10 ± 2	4.0 ± 0.9	1.0 ± 0.1

Three groups of 10 patients (5 F, 5 M; age range 59–73 years) were tested. Each group acted as its own control. Values are means ± standard errors.

Figure 3.5 Effect of changing adjacent rows (Table 5a) to adjacent columns (Table 5b). Comparison of relevant data is easier in Table 5b than in Table 5a. Note that the values are better expressed in millimetres, partly to reduce the number of decimal places but mainly because units representing multiplication by steps of 10^3 or 10^{-3} from the base unit are preferred in the International System of units (SI).

1.58). Convert obsolete units to SI units but don't be over-precise in doing so unless the context and your discipline demand absolute precision. For example, convert 1 gallon to 4.5 litres, not 4.546 litres, in a general context but use exact conversion factors for precise work in the physical sciences.

Remove surplus zeros from very large or small numbers by choosing appropriate factors or units for column headings. Keep numbers to between 0 and 1000 by writing 'mg' instead of 'μg' at the head of a column containing numbers such as 115 000, 192 000 and so on.

Enter zero readings as 0. Use ellipsis marks (...) to indicate that no measurements or observations were made (Huth 1987). Leave a blank if the column or row heading is not applicable. Dashes are ambiguous, as are + and − signs: if you must use these, or if you use abbreviations, explain them in a footnote. Don't use 'ND' or 'NA', which are ambiguous – and unnecessary if you follow the advice just given.

State which test of significance you used (for example, the χ^2, t or F test). Give the probability values (P or p, depending on usage in the journal or your discipline). If it is journal style to do so, mark these values with asterisks for the different degrees of significance. Make it clear whether you measured the standard deviation (SD) or the standard error of the mean (SEM). Always say how many observations were made: '$n = 5$', for example (and see Ch. 5, 'Statistics').

Footnotes/headnotes (explanatory notes)

Put general notes that apply to the whole table first, and do not number them. Include abbreviations among these notes, and explain them the first time they appear in a table – unless they are abbreviations that are acceptable without explanation in the discipline covered by your target journal (see, for example, the list of accepted abbreviations published by the *Biochemical Journal* in its 'Policy of the journal and instructions to authors'). Remember that readers generally look at tables before reading the text; explain the abbreviations even if you also explain them in the text.

Link other footnotes to the parts of the table they refer to by using reference characters approved by the journal and placed against the highest level of organization possible (i.e. against straddle headings and column or stub headings in preference to individual entries). Superscript letters or symbols are preferable to numbers, which can easily be confused with values in the body of the table. Journals often ask for the following symbols to be used, in the order shown, even though few of these symbols appear on ordinary typewriters: *, †, ‡, §, ||, ¶, # (asterisk, dagger, double dagger, section sign, parallel sign, paragraph sign, and the hash or number sign). Use **, ††, etc. if you need more than seven symbols. In the body of the table arrange the footnote reference characters in order from the top down, going from left to right in the first row in which a reference mark is needed, and then downwards and left to right from row to row (see Fig. 3.2). If you are using asterisks for probability values, explain the asterisks after the other footnotes.

You can include short descriptions of your methods in tables if your target journal requests or allows this. These descriptions should appear as footnotes below the body of tables that have titles above them, or as 'headnotes' after the titles, according to journal style. But give as much as possible of the detailed description in the methods section of the text rather than in the explanatory notes; footnotes and headnotes are usually printed in smaller type than the main text and long notes become difficult to read. If you use the same methods for work represented in several tables, put the necessary information in the notes to the first such table; in the other tables refer readers to that table. Do the same for symbols and abbreviations if these are the same in several tables.

FINAL VERSIONS OF TABLES

At the revision stage (Ch. 7), look at all the tables again. If there are many of them, and not much text, do the tables tell a coherent story? If not, why not? Do you need to add one or more tables, or make more measurements to complete the account you are giving of your results? Or do you feel you are overpowering the reader with unnecessary detail and need to remove a few tables at this revision stage? If there are only a few tables, are they nevertheless consistent with the story told in the text?

Before making the final versions of the tables, compare the titles and explanatory notes. Have you worded the titles in a parallel fashion in similar tables? Have you removed unnecessary words and inconsistent statements? Have you used footnote reference characters consistently and linked them to the right explanatory notes? Have you explained the abbreviations? Have you referred to the same substances by the same names in every table in which they are mentioned? Have you rechecked the calculations, especially percentages?

Check, too, that tables are consistent with the text. List entries in the same order as they are discussed in the text, and don't confuse readers by listing four characteristics of an item in a table when you refer to only three characteristics in the text.

TYPING TABLES

When you are sure that everything is as it should be, type the tables in their final form, each on a separate page, double-spaced throughout. Number the tables with arabic numerals (1, 2, 3 ...) unless the journal uses roman numerals (I, II, III ...). Place table titles after the table numbers, either on the same line or below the number, according to journal style. Do not underline the titles, apart from words that are normally italicized, unless the journal prints titles in italics. Do not type any words in capital letters or with initial capital letters unless they are normally written that way – or unless journal style requires capitals.

If a table title is more than one line long, double-space the title. Explanatory notes, whether added to the title or placed below the table, should be double-spaced too, as should the rows of data (the field). Single-spaced or narrow-spaced (1½-spaced) tables are extremely difficult for the editor or production staff to mark up satisfactorily for typesetting. In tables with a large number of rows, leave an extra double-line space after every fifth row, to improve legibility.

If you are short of space, split long words in column headings and put the second part on another line. Put a unit of measure in a straddle or column heading instead of repeating it for every entry in the column(s).

If a table is too long for one page, type 'Continued' at the bottom of the first page and 'Table n, page 2' at the top of a new page, then repeat the column headings on the new page before typing the rest of the table. If a table is too wide to be typed on one page, turned sideways if necessary ('landscaped' or 'broadside'), it will probably be too wide for the journal to print.

Don't use small print anywhere, even if your target journal prints tables in smaller type than the text.

Make sure that all the necessary footnote reference characters are inserted in the table and that each character identifying a footnote appears at least once. For the sequence of footnote reference characters see 'Footnotes/headnotes (explanatory notes)' above.

Type the (double-spaced) explanatory footnotes for a table immediately under that table, not on a separate page unless the table fills the whole page. If necessary, continue the footnotes on a new page (with 'Continued' at the bottom of the first page and 'Table n, page 2' at the top of the second page). Identify each footnote with a reference character and make sure it is linked to the correct matching character in the body of the table.

Don't type or draw more lines above, below, within, or around tables than journal style requires. In particular, don't insert vertical lines between columns unless this is journal style.

Align numbers in the body of a table on the decimal point. If ± or = signs or 'to' ('6 to 8') are used in the columns, align first on these and then on the decimal point:

```
 68.1  ±  1.5
234.0  ± 21.0
  0.29 ±  0.03
```

(See also Ch. 9, 'Numbers and mathematical formulae').

SUMMARY

(1) Decide which tables you need and (re)design them for publication;   (2) Decide whether any tables would be better presented as graphs;   (3) Design simple, comprehensible tables that fit the journal's format easily;   (4) Draft a concise title for each table; (5) Arrange the columns in the most suitable order and write short and simple column headings;   (6) Edit the data in the field; (7) Use appropriate explanatory notes, accurately linked to the parts they refer to;   (8) At the revision stage, confirm your choice of tables for the final paper;   (9) Type the final versions of tables carefully, double-spacing everything, and using ordinary-size type.

CHAPTER FOUR

Preparing effective figures

Choosing and designing preliminary versions   Graphs   Maps
Flow charts   Photographs   Drafting legends   Making final versions   General advice   Size and shape   Plates   Lettering and
symbols   Lines and curves   Axis labels and scale marks   Maps
Correcting mistakes   Checking final legends   Identifying and
posting figures

Figures (illustrations) for papers intended for publication must be
designed with publication and the needs of your readers in mind. The
requirements for figures that will appear in print are different from those
for slides. It is best to design for each medium separately. If you plan
carefully, however, you may be able to use the same artwork for a journal
article, a poster presentation, and slides (see Ch. 12).

Figures and their legends, like tables and their titles, must be independent of the text and of each other. Readers scanning journals often look at
the figures, together with the title, abstract, and tables, before deciding
whether the text of an article is worth reading. It is therefore a good idea
to start preparing figures at an early stage, before you even think of
drafting the text.

Figures are meant to demonstrate evidence vividly. They must be
simple and clear enough for readers to get the message immediately.
Inadequate figures spoil many papers, and few journals now redraw or
reletter figures submitted by authors. Have figures made professionally,
if possible, and treat the cost of professional artwork as an essential part
of the project's budget.

If an artist or photographer who makes figures for you is employed by
your organization, note that copyright in the figures belongs to the
organization and you should be able to use the figures without obtaining
further copyright permission from the maker. If you employ a freelance
artist or photographer, however, the copyright belongs to him or her
unless a written agreement is made transferring copyright and other
rights to you; an agreement of this kind makes you the owner of a 'work
made for hire' (CBE Scientific Illustration Committee 1988, p. 79–82,

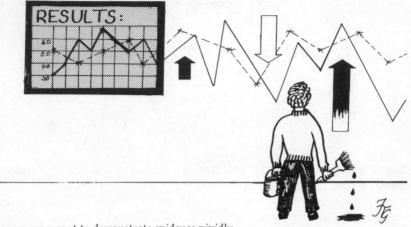

Figures are meant to demonstrate evidence vividly.

251–263). Whoever owns the copyright, an artist who makes anatomical drawings, for example, may sign the drawings if he or she wishes to do so.

If you can't have figures made professionally, at least try to consult a technical artist or photographer about the design (show your adviser a copy of the journal and its instructions to authors). You should also consult a recommended book on how to prepare illustrations, then follow the advice there and in the second half of this chapter. This chapter is itself based mainly on Reynolds & Simmonds (1981) and the Council of Biology Editors' manual, *Illustrating science* (CBE Scientific Illustration Committee 1988). Other recommended books include Cleveland (1985), Schmidt (1983), and Tufte (1983, 1990). If you are using computer graphics, see Simmonds & Reynolds (1989) or, in the earth sciences, Reeves (1989), for example.

CHOOSING AND DESIGNING
PRELIMINARY VERSIONS

First check whether the journal limits the number or size of figures you may submit. Look at any figures you have already made in a preliminary form and decide which ones to use. Design any others you think you will need, then draft legends for all of them. Don't make the figures in their final form until you are sure what that form is going to be. Some kinds of figures are expensive to make and you may want to change your first choices later, or leave some of them out. Make the final versions at the revision stage (Ch. 7), as described in the second half of this chapter.

Figures may be either line drawings (diagrams, graphs, etc.) or photographs from which 'half-tones' can be made (see 'Photographs' below). Journals may accept photographs of line drawings instead of the original drawings, but some journals ban figures that contain both a line drawing and a photograph.

Choose graphs when trends or relationships are more important than exact values, or when meaning needs to be forcefully expressed, or when hidden relationships or trends need to be revealed. Prepare tables rather than graphs when it is important to give precise numbers (p. 20). Draw a distribution map when the location of data is more important than actual values. Design a simple flow chart if you want to present processes, sequences, or systems in an organized way. Construct an algorithm when you want to show the decisions and steps involved in solving a problem or reaching a particular endpoint. Use photographs when it is important to show the actual appearance of something or someone. Keep the number of photographs to a minimum, because they are more expensive to print than line drawings.

Graphs, maps, flow charts and photographs are discussed in the sections that follow and each of these is mentioned again, in more detail where appropriate, in the second half of the chapter ('Making the final versions').

Graphs

Numerical findings may be presented as scattergrams, as graphs with plotted curves demonstrating relationships between two variables (line graphs), or as bar graphs or bar charts showing data for one variable. Histograms use contiguous bars to show the frequency distribution of observations for each class of a variable such as time or weight. Pie charts show the comparative size of component parts effectively.

Line graphs are used for dynamic comparisons, often over time. They are better than vertical bar charts when you want to extrapolate from a trend. If you nevertheless prefer a bar chart for this purpose, use a vertical rather than a horizontal bar chart. Use a horizontal bar chart for comparing proportions or showing where events occur in time. Use bar charts in preference to line graphs when there is no evidence of a continuum between the experimental points or when the findings can be subdivided and compared in different ways.

Design your graphs to show several trends and relationships at once. Don't, however, try to cram too much into one figure – but don't waste space either. Three or four curves should be the maximum in a line graph, especially if the lines cross each other two or three times. When curves must cross, show which lines run where by making them of different thicknesses or different patterns. See Figures 4.1–6 for good and bad ways of presenting graphs.

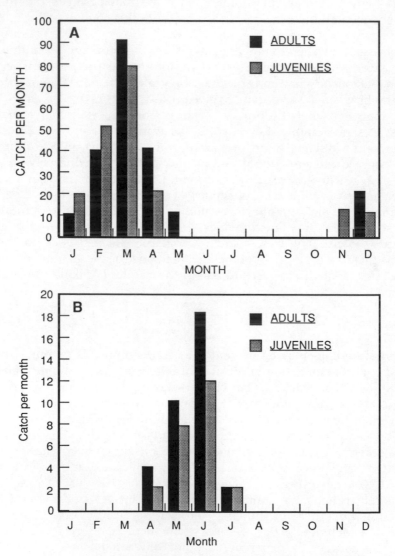

Figure 4.1A, B Graph with some common faults: (1) scales in A and B are different, distorting the information; (2) tint chosen for 'juveniles' reproduces badly; (3) capital letters wrongly used (should be used for abbreviations only); (4) excessive numbering on the y axis; (5) unnecessary use of underlining for the key; (6) unit of measure missing on y axis (7) heavy black frame distracting; (8) calibrations (tick marks) face inward, where they are either unnecessary or may conflict with the data.

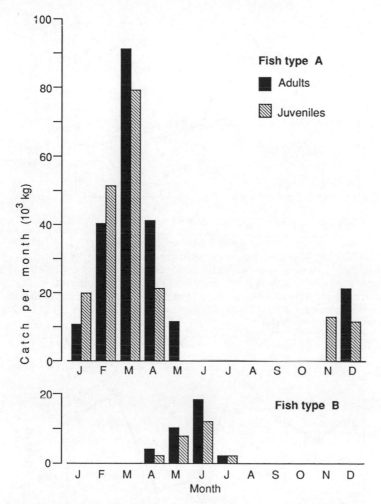

Figure 4.2A, B Clearer presentation of the findings in Figure 4.1, using slightly less space (which editors like). The two graphs are drawn to the same scale, there is economy of line and lettering, the line pattern chosen for 'juveniles' reproduces well, and type sizes are correctly used.

Draw curves as straight lines between data points or as smoothed curves fitted by an equation (smoothed curves won't necessarily pass through every data point).

Place a dependent variable on the vertical axis (the y axis or ordinate [o]). Place an independent variable (such as time, whose values are not affected by changes in other variables) on the horizontal axis (the x axis or abscissa [a]). If you measured two variables in different ways, don't

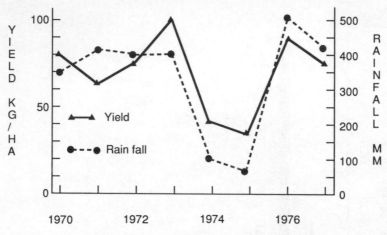

Figure 4.3 A messy-looking graph, with several faults: (1) key has double symbols; (2) lines are joined to the points, and at certain line angles the triangular points disappear; (3) curves are distinguished by type of data point as well as by type of line; (4) vertical lettering is difficult to read; (5) lettering all in capitals produces unrecognizable abbreviations; (6) lettering is placed too far below the *x* axis; (7) no unit given for *x* axis (1970 and so on might not be years); (8) ticks inside axes can be confused with the data.

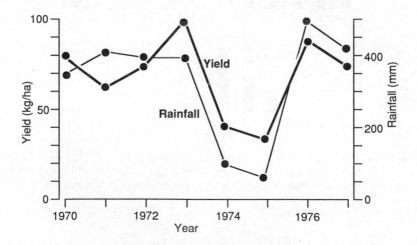

Figure 4.4 Better presentation of the findings in Figure 4.3: (1) no need for key if the lines can be labelled directly; (2) data points stand out better when lines are broken round them (note that on slides data points can be left out if the change of direction at each point is sufficient); (3) curves are distinguished by type of line only; (4) lettering is correctly positioned and arranged to read upwards on both *y* axes; (5) tick marks are placed outside axis. Note that if there is no common zero the horizontal and vertical scales should have separate origins, as shown here.

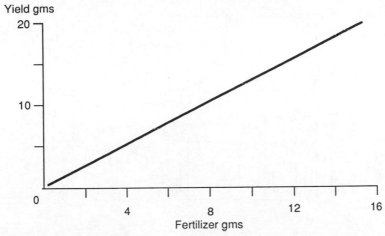

Figure 4.5 Another faulty graph: (1) regression line suggests that yield increases in direct proportion to the amount of fertilizer added, but neither the experimental values nor indications of their accuracy are given; (2) 'gms' is not a recognized abbreviation.

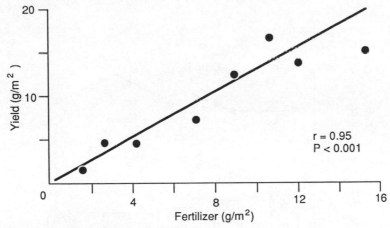

Figure 4.6 Clearer and more informative presentation of Figure 4.5: (1) individual points and regression data are given; (2) units are correctly expressed (in the singular, placed between parentheses); (3) P and r values are included (but may be omitted from slides).

compare them on the same axes (Fig. 4.7). Draw the curves for the two variables separately, using one common axis where appropriate (Fig. 4.8).

Plan graphs so that they need as little lettering as possible. For line graphs, draft short but informative descriptions for the axes (see Figs. 4.4 & 4.6). Use the same symbols when the same entities occur in several

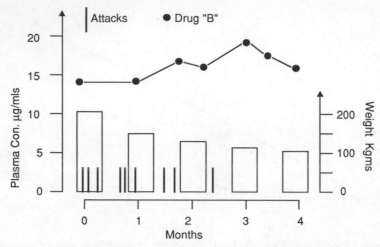

Figure 4.7 A graph in which several sets of data have been superimposed in an attempt to save space: (1) data are measured in unrelated units; (2) labels on the scales are not expressed properly; (3) arrows at the ends of lines are an unnecessary distraction.

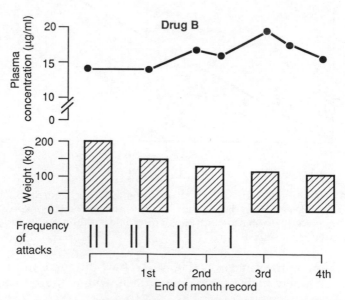

Figure 4.8 Correct method for comparing variables measured in different ways (cf. Fig. 4.7): (1) data are separated into three sets; (2) time scale is more accurately described; (3) space is legitimately saved by a break in the plasma scale (note that zero is retained: omitting it can sometimes lead to misinterpretation).

figures. Use the same coordinates for different figures if values in them are to be compared.

Maps

Maps, like other figures, must carry a message. Decide first what the message is, and state it briefly in the legend.

Maps may be either geographic maps showing large or small areas of the world or thematic maps showing the distribution of various kinds of quantitative or qualitative data. Both kinds of map are best prepared by a professional cartographer but computer-produced maps can also reach acceptable standards for publication. (For detailed information about map-making see, for example, Robinson et al. 1978; see also CBE Scientific Illustration Committee 1988, pp. 116–133, Bryant & Cox 1983, Brouwer 1983.)

Distribution maps are of several kinds. Chloropleth maps use different tones of grey and different kinds of hatching to show areas with different rates, ratios, or frequencies. Point symbol maps use dots, circles, or other symbols for absolute values or frequencies. Isoline maps, familiar from weather charts, use lines joining points of equal value to show the boundaries of variables such as height or rainfall. Flow-line maps use arrows of different length and width to show the flow and amount of a variable.

When you have chosen the kind of map you need, decide how much detail to include and what scale and which kind of projection to use. If you are making a plan of a small area, for example an archaeological site, include the grid references.

For many maps you first need to obtain a suitable base map from a university map library, a geological survey, or a computer graphics program or other collection. Remember to obtain permission to use the map if it is not in the public domain. Then prepare a sketch (a compilation) showing what should go in the printed map. The compilation should include a typed list of everything that needs to be set in type. Arrange this list by categories – cities, rivers, mountains, etc. – and indicate the importance of the various features.

Algorithms and other flow charts

Algorithms take readers on a path through complicated processes or systems by showing where decisions have to be made to arrive at an endpoint or a correct answer. Algorithms can be list-form algorithms, flow charts (logical trees), or MAP (Maintenance Analysis Procedure) layouts (Turk & Kirkman 1989). Another kind of flow chart simply illustrates processes rather than showing where decisions have to be made.

A list-form algorithm consists of numbered statements and questions, with the answers leading readers to the next relevant number. One kind of flow-chart algorithm is shown in Figure 4.9. MAP layouts have the advantage that they can be produced on a typewriter (Fig. 4.10).

Base your choice of which type of algorithm or flow chart to use on what your readers will be using it for. Use standard symbols (e.g. ISO 5807:1985) that are internationally recognizable. Provide a key to symbols when necessary. Use arrows to indicate the direction of flow, where appropriate, and make sure that readers can follow the main movement and the feedback loops easily.

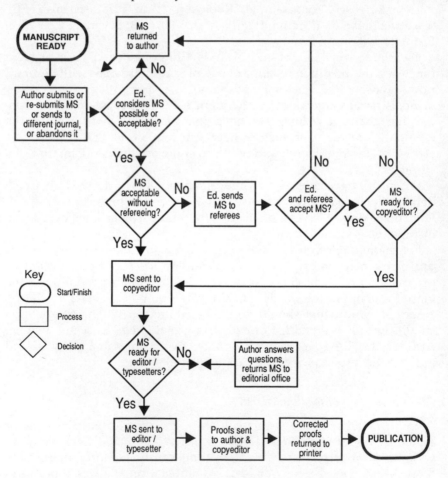

Figure 4.9 An algorithm showing stages on the road to publication.

```
00
Manuscript   ready.
01
Author submits, re-submits or
submits to different journal.
|
MS POSSIBLE OR ACCEPTABLE?
Y    N
|    02
|    - MS returned to author
MS ACCEPTABLE WITHOUT REFEREEING?
Y    N
|    03
|    - MS sent to referees
|    MS ACCEPTED AFTER REFEREEING?
|    Y    N
|    |    04
|    |    GO TO STEP 2
|    - MS READY FOR COPYEDITING?
|    |    Y    N
|    |    |    05
|    |    |    GO TO STEP 2
06
- MS sent to copyeditor
|
MS READY FOR TYPESETTER/EDITOR?
Y    N
|    07
|    - Author answers queries &
|      returns MS to Editorial Office.
08
- MS sent to editor/typesetter.
09
-Proofs sent to author & copy editor
10
- Proofs returned to printer
|
Publication
```

Figure 4.10   A MAP (Maintenance Analysis Procedure) layout giving the same information as Figure 4.9.

Photographs

If you plan to submit photographs of patients, you or the photographer must obtain their informed consent for the photographs to be used. You must also make sure that no one is identifiable – note that masking the eyes is not always enough to mask a person's identity.

   If you decide that it is essential to show the actual appearance of a record, person, or object, choose or prepare photographs of the best possible quality. During the printing stage black-and-white photographs (continuous tones) are usually photographed through a screen with a grid of 133–200 lines to the inch (54–80 lines/cm), to produce the dots that make up the printed 'half-tone' image. The printing process transfers the image from one surface to another several times, usually with some loss of clarity at each transfer, so half-tone figures are rarely better than their originals.

Photographs intended for journal publication should be sharply focused, with a good – but not too wide – range of light and dark tones. If the journal uses high-quality glossy paper and a high-resolution screen for half-tones, the range of contrasts from highlights to shadows can be greater than for the 133 to 150-line screening used in the average journal. See CBE Scientific Illustration Committee (1988, p. 219–233) for a more detailed description of photographs.

Some kinds of records don't come out well in photographs. If an electrophoretogram, autoradiogram, or paper chromatogram is likely to reproduce poorly, if at all, draw arrows to show important points, or make a drawing based on the original record and include both the drawing and a photograph of the original record in the figure (if the journal accepts this kind of illustration). Similarly, it may be more effective to draw a diagram of a piece of equipment or describe it in the text than to include a photograph.

If you are thinking of including colour photographs, find out first whether your target journal accepts them (see the instructions to authors, or consult the editor). Some journals do not print colour photographs; others won't accept them unless the author agrees to pay some or all of the cost of reproduction, which is usually high.

If you submit colour photographs or transparencies from which you want black-and-white prints to be made, there will be some loss of definition in the prints. It is better to keep control of the quality, as far as you can, by having black-and-white prints made in your institution or elsewhere locally. (See p. 156 for more about colour slides.)

Drafting legends

Draft a set of legends (informative titles or captions, with brief explanations) in the style usually used in the journal. Don't plan to put titles on the figures themselves (but see p. 151 if you are making a slide for an oral presentation). Keep the draft legends and the figures in their preliminary form beside you while you write the first draft of the paper (Ch. 5).

The rest of this chapter discusses the preparation of the final versions of the figures and legends. It can best be read (or reread) at the revision stage.

MAKING THE FINAL VERSIONS

General advice

When you make the final versions of the figures at the revision stage (Ch. 7), put each figure on a separate sheet of paper, preferably of the same size as the typing paper used for the text (see 'Size and shape of figures' below).

If you use a computer graphics program for your figures the output must be of good enough quality to be satisfactorily reproduced by conventional printing techniques. Lines must be clearly drawn, with good contrast against a plain white background (see 'Lines and curves' below). Make sure that axes, curves, and lettering are solid – not made up of dots – and smooth, not jagged. Symbols must be clear, large, and distinctive: don't use circles and hexagons on the same figure, for example (see 'Lettering and symbols' below).

A transparent drawing surface with a light below it will be invaluable if you are making your own drawings by hand. Use good-quality heavy white paper for line drawings – art-coated proofing paper is recommended by Reynolds & Simmonds (1981). Or use white card or graph paper with faint lines of a colour that doesn't reproduce when photographed and of a type that doesn't repel ink. Avoid using tracing paper because it tears easily and may have a slightly greasy finish.

Use technical drawing pens and black drafting ink, well stirred, or use a drawing pen with an ink reservoir. Good quality pens are made to the micronorm standards for draftsmen (ISO 9175-1 & 2:1988) and produce lines of standard widths. The preferred line widths for most graphs are 0.35, 0.5, 0.7, and 1.0 mm. Hold pens with reservoirs at a 90-degree angle to the paper surface. Draw slowly, at a steady speed, with a light but positive pressure. If you are filling in solid black areas, use a technical pen or a paintbrush, never a ballpoint or felt-tip pen. Or use preprinted sheets (tone sheets) containing solid black shapes which can be cut and stuck in position easily. You can also use tone sheets for hatching, dots, and various shapes.

Make it clear what the labels on figures refer to but don't draw a confusing number of lines and arrows. Include scale bars on maps, micrographs, and anatomical drawings.

Make well-focused photographs of hand-drawn line drawings to submit to the journal (computer drawings don't need to be photographed). Before the drawings are photographed make sure that every detail on the originals is correct, including spelling, and that abbreviations and symbols are approved by the journal or by your discipline. Keep abbreviations to a minimum: write words in full if there is enough space to do so without cluttering up the figure.

43

Submit glossy photographs unless the journal specifies matt prints. Protect each print with an overlay – a sheet of paper taped to the back of the print and folded over the front. Don't mount photographs on paper or card unless you are grouping them on a plate (see 'Plates' below) or unless the journal asks you to mount them in this way.

Size and shape of figures

For preference, make the original figures fit well within the limits of the page size on which the text will be typed – either A4 (about 210× 297 mm), or 8.5×11 inches (about 216×279 mm). Illustrations of typing-paper size are easier to handle than very large figures and are less likely to get damaged in transit or in the editorial office. Very small figures risk being lost at some stage of the production process. To prevent this, tape them lightly to a sheet of typing-paper, or put each one in a transparent envelope and tape each envelope lightly to a sheet of paper.

Many figures are reduced during printing to a size that conveniently fits a column or page of the journal. Make original line drawings not more than twice the final size, which you can estimate by measuring similar kinds of figures in the journal. Line drawings are most conveniently drawn at twice the final size within a 130×200 mm (5×8 inch) rectangle on an A4 or 8.5×11 inch sheet of paper or card. Make lines and curves twice the final thickness too (see 'Lines and curves' below).

If several of your line drawings show similar data, draw them all to the same scale (compare Figs. 4.1 & 4.2). If you make your figures twice the width of a column or page of the journal (allowing for a 50% reduction), the height will then be obvious. Note that if a figure 100 mm wide×100 mm high is reduced by 50% (to 50×50 mm), the area of the figure is reduced by 75%, not 50%.

A ratio of 2:3 for width:height is usually best for figures in journal articles, but the demands of the subject come first. If you need to include a figure that must not be reduced and that is likely to take up most of a printed page, try to leave enough space for the legend to be printed below it. Avoid submitting figures that will be larger, after reduction, than a journal page. Be economical with space; arrange labels, for example, in such a way that figures are as compact as possible.

Photographs are best made at either final size or one-fifth larger than the final size. If several photographs are likely to be printed close together on the journal page, make them match each other in either width or height, if not in both dimensions (and see 'Plates' below).

To avoid the need for large photographs to be reduced so much that essential details disappear, decide which parts readers must see, then cut off or mask the rest. Most photographs can stand being 'cropped' in this

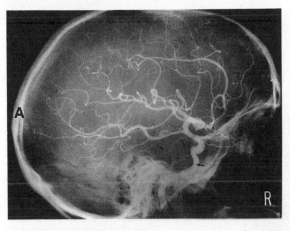

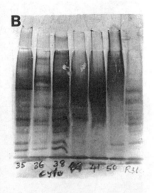

**IMMUNOBLOTS OF NORMAL PLATELETS
IN 7.5% SDS-PAGE GELS**

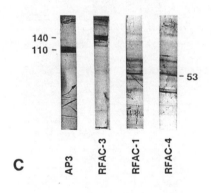

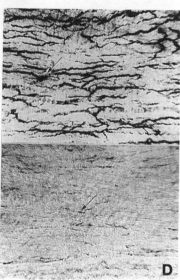

Figure 4.11 A badly prepared plate: (1) photographs are cut carelessly with scissors and do not form a rectangle; (2) A, B, C, D labels are inconsistently placed and are too small; (3) arrows on A and D are almost invisible; (4) lettering on B is as originally scribbled on the gel; (5) material to be photographed for continuous tone prints should not include lettering (the result is either rough lettering or poor quality tone, because photographic film cannot be equally sensitive to high contrast images, such as lettering, and continuous tone images); (6) D has no scale (scales are essential on photomicrographs because they reduce in the same proportion as the image). (Photographs by courtesy of the Medical Illustration Department, Royal Free Hospital, London.)

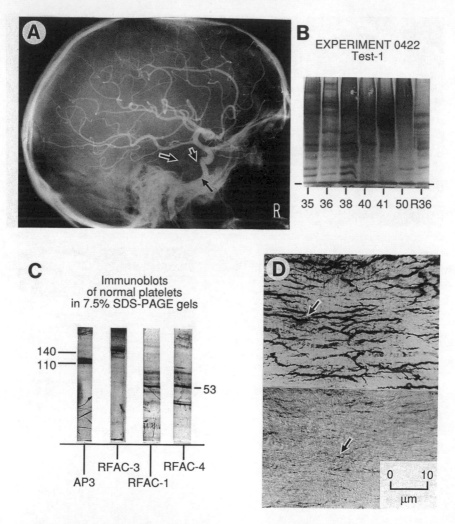

Figure 4.12 A better layout of the same photographs as in Figure 4.11: (1) photographs are properly trimmed to make a rectangle; (2) labels A, B, C, D are consistently placed (and placing one self-adhesive circle on top of another makes the background to lettering look whiter); (3) arrows in A and D are large enough and are outlined in white (black arrow placed on top of a larger white arrow); (4) lettering for B and C is drawn separately from the photographs, and lettering and photographs are then pasted together for the printer (separate treatment gives a better result); (5) D has a suitable scale; (6) lettering is large enough to reproduce at half its original size. (Photographs by courtesy of the Medical Illustration Department, Royal Free Hospital, London.)

way. If you can't cut the original or mask it satisfactorily, mark the edges or back of the prints, or mark a photocopy, to show the journal's production staff which area is to be used.

If your photographs show a mixture of coarse and fine detail, mark a photocopy to show the finest detail that must remain visible in the printed figure (in a photograph of fish, for example, are you concerned with the overall body shape or the arrangement of the smallest scales?). Draw the production editor's attention to any subtlety that must remain visible (for example, clouds of fine particles coming from behind the gill-covers of a feeding fish). That is, say exactly what is important in the photograph. Don't rely on the legend to do this: figures and legends are dealt with separately during production of the journal.

Plates

Plates are inserted glossy pages of half-tone illustrations and are a relic of the days when special paper was needed for engravings or half-tone illustrations. Because the paper used for the text is now usually adequate for most photographs the need for plates has almost disappeared. Some journals, however, still include plates, and some editors call any page filled with figures a 'plate'.

If you want several photographs to form a single plate or fill a page, provide a photocopy showing how the photographs should be arranged, or mount the photographs on white card. Don't allow varying amounts of the card to show between the photographs (Fig. 4.11) and don't paste dividing strips between the photographs. If any divisions are needed, the production staff of the journal will mark them. Make sure that all the corners of the prints are right angles and that the outside edges form a rectangle (Fig. 4.12). Remember that the prints will all be reduced to the same degree, so ensure that fine detail will not be lost.

Letter or number the photographs making up the plate and arrange them in order from left to right and top to bottom. Arrange them in such a way that the whole assembly is in proportion to a column or page in the journal. Match the size of the assembly to the width of the journal page if it cannot be matched to both width and height, as discussed above ('Size and shape of figures').

Lettering and symbols

The journal's instructions should tell you whether to put lettering and other symbols on the original illustration or whether to pencil them on a transparent overlay or write them in non-reproducing crayon on a line drawing as a guide for the publisher's artist. If an overlay with lettering on it is required, attach the overlay firmly to the back of the figure so that it won't move and will show the position of the lettering accurately.

If you are responsible for the lettering, make sure that it looks professional. Many good figures are spoilt by messy or uneven lettering. Before you make the final figure, practise with whatever lettering method you are using.

Make letters and symbols large enough to be legible after reduction (see p. 50). The final height of an 'x', for example, should be at least 1.5 mm on line drawings or 2.5–3.0 mm on photographs (CBE Style Manual Committee 1983, p. 73). Don't use heavy black lettering (bold lettering), which may look too dark when printed (Fig. 4.3). If you use preprinted lettering, choose a simple face such as Univers or Helvetica (Helvetica is used in the figures in this chapter). You may need to use italic (sloping) lettering for species names, for example, if the journal requires it, but underlining may be acceptable.

Use the symbols and abbreviations approved by the journal (see 'Lines and curves' below for more about symbols). Place a zero in front of a decimal point (0.1, not .1) and use SI units correctly. Use a solidus or negative superscripts (kg/ha or kg ha^{-1}) according to the journal's custom. When in doubt, use negative superscripts.

In algorithms use standard symbols (ISO 5807:1985) for the various stages, e.g. rectangles for actions or processes and diamonds for decision points (Fig.4.9). Where necessary, use arrowed lines to show the direction of flow in algorithms and other flow charts. Write simple questions or sentences inside or beside the different shapes, as needed. Remember that the lettering will be reduced in size during the production process.

On photographs with a dark background use letters, numbers, or symbols preprinted on a white background, or use a stencil to draw these characters on white labels (Fig. 4.11). To prevent the background showing through when you use preprinted characters, use two such labels, one on top of the other.

Use scientific stencils with raised edges (ISO 9178-1:1989, ISO 9178-3:1989) and technical drawing pens for drawing curves, lines, chemical structures, and mathematical symbols. Alternatively, use preprinted shapes or letters, or use an acceptable computer graphics program (see p. 43). Never draw letters freehand unless you can do this to a high standard. Never change 'u' to 'μ' by hand or make similar changes on an original figure. Be sure to use the same style of lettering throughout, except when you are distinguishing different features by using different styles.

Leave a space between letters of about the same width as the width of the strokes in the characters. The space between words should be about the width of the letter e. The space between the bottom of an 'x' on one line and the bottom of an 'x' on the next line should be at least 1.5 times the height of a capital letter in the size of lettering you are using.

Lettering produced on a clear adhesive strip with a lettering machine is faster and easier to use than dry-transfer lettering. This kind of lettering machine is expensive but is a sensible investment for an institution. Make sure that the device you use produces a range of sizes and spacing suitable for your purposes.

If you use dry-transfer lettering, draw light guidelines for the lettering in non-reproducing blue or green pencil or crayon (test for reproducibility by making a photocopy of a drawing, or ask your supplier for these pencils or crayons). Move a burnishing tool or the point of a ballpoint pen lightly backwards and forwards across each character to transfer it to the artwork below. Put an empty part of the sheet of letters over each line or word when it is finished and burnish the lettering by rubbing strongly with the flatter end of the burnisher or the smooth top of a suitable pen, to make the letters stick firmly to the drawing-paper.

If you are using preprinted hatching or other tones, rub the sheet down carefully and fill the required area precisely. Changing the direction of burnishing occasionally will help to cover the area properly. Never put tone sheets on top of self-adhesive tapes if you use these tapes for lines or curves on the artwork. Remove air bubbles between tone sheets and the drawing paper by reapplying the tone sheet slowly and carefully (use the kind of tone sheet which is easy to remove before it has been burnished), or prick small bubbles with a scalpel or pin, and then burnish the area. Put tones around lettering, not on top of it (cut a window for the lettering).

Dots in tones should measure 0.5 mm, with 0.5 mm between dots. Parallel lines or cross-hatching should be 0.35 mm wide, with a minimum of 1 mm space between the lines. If you need to use several tones, they should be easily distinguishable. Don't put dark tones in the upper part of a bar that is blank or lightly hatched in the lower part. Avoid putting different tones on top of each other: dirt that is difficult to remove can easily be trapped between the sheets, and moiré patterns can occur that will not print well.

Lines and curves

Do not draw rectangles around line drawings (Fig. 4.1) unless the journal specifically requires this. Make curves and plotted points bolder than the axis lines unless there are a lot of curves or points. Make plotted points stand out well. If they fall on an axis line, break the axis on each side of the point. (Note that the baseline of a bar graph can often be left out.)

If you extrapolate a line or a curve beyond the observed points, distinguish the extrapolated part from the rest – use a dashed line if the rest is continuous, for example.

Differentiate curves by using different symbols for points joined by the same type of line (– ● – , – ○ –) or by using the same symbol joined by different types of line (Fig. 4.4). Using different types of line is preferable when only two curves need to be differentiated. If the symbols are clear, there is no need to overdo the differentiation by using both methods together (Fig. 4.3).

For experimental points in line drawings use the journal's preferred symbols – usually ●, ○, ▲, △, ■, □ – and make them two or three times the width of curves in the graph. If data points overlap, draw them overlapping. If the points coincide, one solution is to use just one symbol for these points (CBE Scientific Illustration Committee 1988, p. 96). Don't put circles next to squares: it is difficult to distinguish these symbols after reduction. Don't use tiny hexagons or X, +, ☉, or *: these symbols are not distinctive enough or do not reproduce well. Make sure that all data points are clear, especially if they are produced by a computer.

Where appropriate, draw a vertical line to show the standard deviation or the standard error of the mean for each data point. These lines are usually drawn in pairs, one above and one below a data point or the end of a bar, but it is really only necessary to draw the top line of each pair. Make the lines thinner than other lines in the main body of the data – for example, 0.35 mm thick if you are using 0.5 mm for the curves – and don't let them overlap. In the legend, tell readers what the vertical line represents and always say how many observations each mean is based on.

Make all curves and lettering suitable for reduction by 50%. If an original figure larger than a sheet of typing paper is unavoidable, calculate the size of the lettering and the thickness and spacing of lines carefully. Details that are too fine won't register when the original is reduced, and cross-hatching and other lines that are drawn too heavily may come out solid black. In ordinary print the finest line that can appear is about 0.16 mm thick, or 0.2 mm for a white line on a dark background. Finer lines or lines that are not properly inked will not show up clearly. If lines as thin as 0.16 mm are to be printed satisfactorily the original, drawn at about twice the size (0.35 mm is suitable), must be perfect. Make a photocopy of the figure at the same size as the intended final size to see whether it will reproduce well enough.

Use thinner lines for hatching and other kinds of shading than for the rest of the drawing (see 'Lettering and symbols' above). If you want broken lines, draw continuous lines first, then break them carefully with white paint or correcting fluid, or, on art-coated paper, use a scalpel to scrape gaps in the lines. Alternatively, use dry-transfer or other pre-printed lines.

Label curves and so on directly, where possible, rather than using a key (but check your target journal's practice). Keep labels well away from lines and curves and position them horizontally. If a key with unusual symbols is needed, put it in the body of the figure rather than in the legend (unless the journal gives other instructions), because the typesetter may not be able to reproduce the symbols satisfactorily in the legend. Make sure that key symbols in the body of the figure can't be confused with plotted points, but don't enclose keys in a box. If the text is to be printed in more than one language (more likely for a book than a journal), number the key symbols in the body of the figures and use numbered keys in the legends.

Axis labels and scale marks

Label the vertical axes of graphs as briefly and simply as possible. If labels for left or right y axes cannot be written horizontally, write vertical labels parallel to the axis, for reading upwards. Centre the x axis label below the horizontal line. If centring the label in a hand-drawn graph is difficult, start it below the first scale point numeral.

Use the axis labels to show the variable and the unit of measurement only, rather than including the subject matter of the graph. Name the variable and give the units (with SI abbreviations, usually in lower-case lettering) in parentheses. Don't use ambiguous multipliers such as $\times 10^3$ on axis labels (see p. 25). For the variables use lower-case lettering, reserving capitals for the first letter of the first word, for any other words usually written with an initial capital, and for abbreviations and acronyms. Explain abbreviations in the legends.

Choose scales for axes carefully. If an axis does not start at zero (or 1 for log scales), mark a break in the scale. Show clearly where a change in scale occurs. Don't extend axis lines beyond the last marked scale point, and don't end them with an arrow pointing away from zero.

Mark scale calibrations (tick marks) clearly. Put them outside the axis and mark and number only as many as are necessary for clarity.

Always include scale bars on micrographs. This is preferable to giving magnifications in legends because the scales will remain correct if the figure is reduced during the production process or is later printed at some other size. Alternatively, if you want to state the original magnification, give it in the legend and add the photographic reduction when it is known – e.g. '$\times 12\,000$, photographic reproduction at 80%' (where 'at 80%' means the figure is 80% of its original size – and 64% of its original area).

51

Maps

Draw maps double the size they will be when printed in the journal and plan to make lettering, lines, and symbols large enough to stand this 50% linear reduction. If you are using a computer graphics program to make your maps, make sure that the output is of a suitable standard for publication.

Include a scale bar and show the north point if it is not at the top. Name the projection used and indicate the latitude and longitude or the grid reference, as appropriate.

Spell place names, especially transliterated names, in accepted ways. Make sure that all geographical names and features mentioned in the text are also found in the map, but don't overload the map with irrelevant information. Marking one or a few well-known major features (rivers, cities) may be enough to orientate readers. Write the names of different features in different sizes or styles of lettering. If you write names out of the horizontal, make all the lettering run in approximately the same direction (Butcher 1983).

Make sure that lines will not disappear on reduction (see 'Lettering and symbols' above). Place keys for symbols or patterns within the map itself.

Correcting mistakes in drawing

Correct small mistakes by scraping them gently with a scalpel, if you are using art-coated paper. Or use white paint or correcting fluid, but don't use too much fluid because wrinkles or bumps may appear that will attract dust or stop tone sheets being properly applied.

Correct larger areas by sticking a patch of white paper over them, using paper of the same kind as for the rest of the drawing. Trim the patch as close as possible to the edges of the drawing. Put adhesive on the patch, not on the artwork, and don't use too much adhesive. If surplus adhesive escapes from under the patch, remove the adhesive with a scalpel. Don't use adhesive tape on a patch. Don't patch drawings made on tracing paper or film.

Try not to put a patch where newly drawn ink lines will need to cross the edge of the patch. If you can't avoid drawing lines from a patch to the original drawing surface, make sure that the edge of the patch is firmly stuck down. Remove any adhesive that has escaped and draw the new line from the patch outwards. If there are any small gaps in newly drawn lines, fill them in with a finer pen than you used for drawing the lines. If any blobs of ink appear, let the ink dry and then trim the blobs to the correct line width with a sharp scalpel, holding the blade at an angle of about 90 degrees to the surface of the paper. If you damage the surface of the paper when you scrape it to make corrections, rub the surface with a burnisher before drawing any more lines on the paper, because the ink may spread like a blot.

Legends: a final check

Check the draft legends against the final versions of the figures and rewrite them, if necessary. If you have renumbered the figures since drafting the first version of the legends, renumber the legends and see that you have explained all symbols and abbreviations if these aren't identified within the figures. Compare the wording of the legends; use parallel wording for similar illustrations and remove unnecessary and inconsistent statements. Avoid using 'see text for explanation'.

Type the legends as a consecutive series on one or more pages, as needed. Begin each legend 'Fig. n' or 'Figure n', according to the journal's style. Type the title without underlining, unless the journal prints figure titles in italics. Don't use bold-face (heavy) type.

The descriptive material in a legend may include letters or symbols that have to be put in by hand or that are represented by typesetting codes. Check that these letters or symbols correspond to those on the figure itself and that any typesetting codes are correct. Make sure that spelling, abbreviations, and symbols are consistent between each legend and its figure, and consistent between legends. Make sure that they are consistent with the text, too.

Identifying figures and preparing them for posting

When the final versions of the figures are ready, arrange them in the order in which you refer to them in the text. Number photographs and line drawings as a single series unless the journal prints photographs as a separate series of plates. Do not write the numbers on an area of the figure that will be printed unless you are arranging the figures on a plate (see 'Plates' above). Instead, write the figure number, your name or the first author's name, and the short title of the paper on the back of each figure, preferably on a label. Some journals may ask for the identifying labels to be put on the front of the figure but outside the area to be printed. Show which side of a figure is the top by writing 'Top' in the appropriate place.

If you put figure numbers directly on the back of a photograph instead of on a label, write very lightly near the top edge of the photograph. Use a soft pencil, never a felt-tipped or ballpoint pen – the marks these pens make can show up on the printed figure.

Never put paperclips on any illustrations, whether line drawings or photographs; the marks they make may also show up on the printed figure.

Don't mount figures on card unless the journal asks you to do so.

When your article is ready to post, put the figures flat between two sheets of stiff card slightly wider and longer than the figures themselves. Then put card and figures in a strong envelope that fits the card, with the manuscript and your covering letter to the editor (see Ch. 10, 'Mailing the manuscript').

SUMMARY

(1) Decide which figures you need and design preliminary versions; (2) Draft a set of legends; (3) Make the final versions of the figures to fit the journal column or page; (4) Letter the figures to professional standards; (5) Make lines and curves clear, and label and differentiate them clearly; (6) Label axes simply and clearly, and mark scale calibrations clearly; (7) Correct mistakes in drawings carefully; (8) Check the draft legends against the final versions of the figures; (9) Number and identify the figures and prepare them for posting.

CHAPTER FIVE

Writing the first draft

Practical preparations for drafting   Place and time   Writing
materials   Form of output   Getting started   Style and grammar
Translations   Nomenclature and abbreviations   References
Headings   Drafting the body of the paper   Introduction
Materials and methods   Results   Making a preliminary presentation   Discussion   Title   Abstract   Key words   Acknowledgements   Appendix   List of contents   Burying the first draft

When the preliminaries are completed and the tables and figures have
been chosen and designed, start drafting the text. If possible, write the
draft before you dismantle apparatus, dispose of organisms, or take
some other irrevocable step.

This chapter discusses practical preparations for writing and then
describes what goes into the main sections of a paper, starting with the
introduction. The final title and the abstract, which of course come first
when the paper is printed, are discussed after the main sections of the
paper, because all the main sections must be ready before you write these
two parts.

PRACTICAL PREPARATIONS FOR WRITING

Place and time

When you start writing the draft, or your share of it, you should, ideally,
cut yourself off from the outside world. Try to find a time when you can
remain undisturbed for several hours and a place where no one will
interrupt you. Write at the time of day when you feel freshest and most
alert. Try not to choose a time when you are physically or mentally tired,
even if that time is convenient for other reasons. Be selfish: these few
hours are the culmination of long and expensive research and you are
entitled to suitable working conditions. Lock the door, unplug the
telephone, ban other potential distractions . . .

In real life, of course, you may have to work while chaos rages around
you, but it is worth trying to remove or escape from as many distractions

55

When you start writing . . . cut yourself off from the outside world.

and interruptions as possible. After that it is a matter of staying in your seat and writing for as long as possible – so find the most comfortable chair possible and adjust it properly.

Writing materials

Whenever and wherever you manage to write, take the outline(s) (Ch. 2, 'Constructing outlines') and all your notes, draft tables, and draft figures with you. You'll also need your laboratory notebooks (if available), a dictionary, reference books, the journal's instructions to authors, plenty of your preferred writing materials, and plenty of coffee or whatever else keeps you alert . . .

Your writing materials may be pen and paper, or a typewriter or computer, or a tape or cassette recorder. If you are thinking of dictating a draft from notes, remember that dictation needs a lot of experience or a lot of correction, or both. Word processing is the easiest method to use but handwriting may be more suitable for first drafts. Word processing can seduce you into thinking you have written a masterpiece when all you have done is written a few paragraphs and moved them around a few times.

If you use word processing, correct your drafts on paper, not on the screen: in making handwritten corrections you are more likely to see discrepancies you might not notice on the screen. Remember to make back-up copies during and at the end of every working session. Losing several hours' work to operator error or to spikes, surges, or breaks in the electricity supply is no fun.

With word processing it is also worth checking the journal's instructions to authors to see whether and how authors' disks or magnetic tapes ('computerscripts') will be used in the production process. If computerscripts are accepted, sort out questions of compatibility with the journal before you begin typing. Programs such as Microsoft Word, WordPerfect and WordStar are often acceptable but if you use a less well-known word-processing program you may have to convert your computerscript into ASCII format – plain text without any word processor codes for bold, italics, underlining, subscript, superscript, or unusual or accented characters – so that text can be more easily transferred to the publisher's system.

Form of output

Whichever method of writing you choose, leave wide margins and at least one empty line between the handwritten or printed lines of the first draft, to allow room for later changes if you make these on paper, as recommended above, rather than on a screen.

GETTING STARTED

If you find it difficult to start writing on the blank page or screen in front of you, leave the introduction for later and start with any section you have already drafted or made detailed notes about. The materials and methods section is often the easiest place to begin, and the results section the next easiest. Once you get going, write as quickly as you can. If the article is short, try to finish it in one sitting, to give it as much unity as possible.

Follow your outline as closely as you can. Most of the paragraphs in the paper should discuss one of the points listed in the topic outline and be introduced by a sentence from the sentence outline, if you prepared one. But you should of course add other relevant paragraphs on topics that come to mind as you write. You should also leave out topics if you realize while you are writing that they are irrelevant. Use the outline as an aid to writing, not as a straitjacket.

STYLE AND GRAMMAR

Don't worry about your writing style or about grammar at this stage. Just write as simply as you can, keeping your intended readers in mind. You can polish the style and correct any grammatical errors at the revision stage (see Ch. 8). Remember that if the paper is aimed at a wide audience many readers will not be specialists in your discipline, and many will not be reading their native language. Writing simply may be easier if you imagine you are describing your work to a friend in a different discipline. Keep reminding yourself of this, and remember that listeners and readers need a story line: a beginning, a middle, and an end, with clear links between each step in the story.

As well as writing simply, try to write interestingly. Most scientific articles have to be serious but that doesn't mean they have to be dull. Let your enthusiasm for your subject show through, at least occasionally: if you want to write that your frightened sea urchins stampeded (Paine & Vadas 1969), do so. Don't expect your readers to agree if you simply state 'This is an interesting finding'.

TRANSLATIONS

If English is to be the language of publication and it is not your first language, choose one of the following methods:

(1) Draft the paper in English to the best of your ability (the best method).
(2) Write the first draft in your own language and translate it into English yourself.
(3) Employ a professional translator who is familiar with the terminology of your branch of science. Include the translation fee when you apply for a research grant, if the funding body allows this.

Persuade a scientific colleague or a correspondent whose native language is English, and whose English is up to date, to comment on the resulting manuscript. If you can't find such an adviser, appeal to the editor for help when you submit the manuscript – but do this only as a last resort. Referees and editors will gladly correct minor mistakes but the English must first be good enough for your meaning to be clear.

NOMENCLATURE AND ABBREVIATIONS

Use the accepted conventions of your discipline for nomenclature, abbreviations, and particular ways of presenting information. If you are not sure what the conventions are, don't worry about them at this stage; instead, check them when you revise the manuscript (see Ch. 7).

REFERENCES

Regardless of the journal's reference style, use the name-and-year system (the so-called Harvard system) for references in the draft text. That is, write '(Smith 1985, Braun et al. 1986)' or 'As Smith (1985) and Braun et al. (1986) report ... '. It is much easier to compile the final reference list from names than from numbers, and if the list is a numbered list the right names are more likely to be attached to the numbers. Alternatively, if you are using a word processor but lack a bibliographic program for managing references, flag the reference sites with a character that is unlikely to appear elsewhere in the text, such as $ or @; later you can search for this character and insert the citations. (See Ch. 6 for detailed advice on recording, choosing, and listing references.)

HEADINGS

Put headings in the text wherever these are likely to help readers to see the line of argument, especially if the paper is long. The main (first-order) headings may simply be Introduction, Materials and Methods, Results, and Discussion (see p. 13). Topics from the topic outline often provide suitable subheadings when these are needed in long papers. Try not to use more than three or four levels of heading: after third-order or fourth-order headings the ranking system is difficult for the typographer to differentiate and for readers to grasp. Make the levels of headings clear to yourself by writing '1' or 'A' in the margin beside main headings in the draft, and '2', '3' or 'B', 'C' beside subheadings and sub-subheadings. You can also keep the levels clear by, for example, putting the main heading in the centre, the second level at the left margin, and the third level a few spaces in from the left margin (but change this layout for the final manuscript, as necessary). Don't use decimal headings (1.5.2.1) unless the journal uses them; they are hard for readers to grasp and they look pretentious. Keep the headings short – one to ten words should usually be enough.

DRAFTING THE BODY OF THE PAPER

Certain questions have to be answered in each main section of a conventional paper intended for a scientific journal. Even if you have decided on unconventional main sections, or are writing something other than a research paper, you will probably need to answer these questions – why?, what for?, how?, where?, what did you find?, what does it mean? – somewhere in your paper.

The introduction

The introduction should persuade readers to go on reading once their attention has been attracted by the title, abstract, tables, and figures. The main questions it answers are:

Why did you do the work?
What is its purpose?

Deal with these questions briefly but as interestingly and as simply as possible. Keep non-specialist readers in mind but don't talk down to them by explaining too much.

Say as concisely as possible what is already known about the subject of your investigation, giving references to the most important publications. Then say why you are querying or hoping to extend what is already known or believed. Provide support for your argument. State the question or hypothesis that arises from your assessment of previously accepted work. Tell readers briefly what you examined, and, if relevant, say where you did the work. Indicate your experimental approach. Point out what is new and important about your work. When appropriate, you should also state the answer(s) you found. You may be telling a story, but it need not be a detective story where nothing is revealed until the last page.

If your paper is about a new method or apparatus, start by saying what is already known or in use, mention its problems or limitations, and state what the new method or apparatus is and what its advantages and disadvantages are (Zeiger 1991).

Don't try to mention everything already written on your subject. One to three paragraphs should be enough for most journal articles. A long introduction is desirable only in a review article or a thesis. If the first draft of your introduction is too long, it probably contains general statements that don't advance the story you should be telling. But don't let length worry you too much at the drafting stage: getting the generalities out of the way may help you to get going with the writing. You can remove the redundant parts at the revision stage.

This example, taken from a report limited to 600 words and five references, shows what can be achieved in a few sentences:

Hay fever has been described as a 'post-industrial revolution epidemic' (1). Successive morbidity surveys in British general practices suggest that its prevalence continued to increase from 1956 to 1982 (2). There is also evidence of a recent increase in the prevalence of asthma (2) and childhood eczema (3). This analysis of data from the National Child Development Study in the UK suggests that the increased prevalence of these atopic diseases may be related to the reduced opportunities for cross-infection in today's families. [Reproduced, with minor modifications, from Strachan 1989.]

Materials and methods

In the materials and methods section or sections – sometimes called the experimental section – the questions to be answered are:

What materials did you use?
How did you use them?

You must also explain why you chose any methods, including statistical methods, to which reasonable alternatives exist.

Describe the experimental design or theoretical approach first – unless the introduction has already made this clear. Say what assumptions you made in working out the experimental design (but lengthy assumptions may be more appropriately dealt with in the discussion instead). Then describe what you did and how you did it, giving enough information for experienced workers to repeat the experiments or assess how reliable the methods are. Include references to accounts of methods published in widely accessible journals instead of repeating details of those methods. Specialists won't need detailed descriptions of materials or procedures, and non-specialists are unlikely to read the materials and methods section with a view to repeating the work. Instead, say 'Growth was measured and analysed according to Braun's method (1988)' (and see p. 62).

MATERIALS

Define the materials you used as precisely as you can without inhibiting the flow of your writing. If the recommendations below are difficult to follow while you are writing the draft, deal with them at the revision stage.

If you studied a particular geographical area, describe it exactly. Use accepted names and spellings and follow recognized authorities for geological nomenclature. If the area is remote or the study very detailed, give a map reference, preferably to a standard atlas.

Give chemical substances, including drugs, their systematic names or recommended international non-proprietary names (generic names).

Don't use trade or local names unless using them is the most accurate way to identify substances – which it may be if substances have not been rigorously purified. If you use a trade name, write it with an initial capital and add the full chemical name, if known, when you first mention the substance. Some journals ask authors to include the names and short addresses (town, province, country) of the manufacturers of drugs with registered trade names. Give the systematic name of enzymes and their Enzyme Commission numbers (Webb 1984) when you first mention enzymes, and afterwards use the trivial names. Give the recognized name of restriction enzymes and the source from which you obtained them.

Give the full taxonomic identification of animals, plants, or micro-organisms, including the source, strain, breed, cultivar, or line, as appropriate. Say which authority is responsible for the nomenclature you use. Give the age and sex of animals, mention their genetic, physio-logical, and dietary status, and describe the conditions in which they were kept. Provide comparable information about plants and micro-organisms.

If human beings or other higher animals were the subjects of your study, give all the relevant information – including age, sex, and state of health. Mention race only if it is relevant. Name the ethics committee or other authority whose approval you obtained for the experiments.

METHODS

Describe the methods in a logical order, including the sequence of procedures for each method. (A flow chart may be useful for illustrating this sequence.) Usually it is best to describe methods in the order in which you used them but sometimes you may want to describe different techniques according to the quantity of the material or the importance of the method. Whichever order you choose, make sure that readers can follow why you did what you did.

Explain the purpose of any procedures whose function is not obvious and provide any background information that may be necessary. If you are studying humans, describe the criteria for including or excluding individuals and, if relevant, your selection method for the sample. Scientific research aims to extrapolate conclusions drawn from a sample to a wider population, so the characteristics of the sample must be stated.

Instead of repeating published details, name the principles on which well-known or previously published methods are based. Cite the original publications, or recent textbooks or handbooks, for all except those few methods that are so well known they don't need a reference. Give the essential features of less well-known methods and cite a publication that gives full details. If you made changes to published methods, describe the changes in detail. If you used a new method describe it in full, then test your description by asking a colleague to do an experiment using the method as you describe it (Booth 1985).

If the methods you use have to be validated by preliminary experiments, present the results of those experiments in the methods section, or in a separate publication. If validation of the methods is the main object of the study, however, put the results of the preliminary experiments in the results section.

Say which methods of statistical analysis you used and make it clear which method was used in each part of your investigation. If you used a large number of statistical methods, name each one again in the appropriate place in the results section. You need not describe well-known statistical techniques in detail but you must say which form you used if a technique (such as the t test) has more than one form. Explain any complex statistical methods you used and why you chose them (Altman et al. 1983). If you used a computer program to analyse your findings, say which program and which version it was, as well as identifying the specific methods of analysis you chose. (See 'Results' section for more on statistics.)

If you did experiments on human beings or other higher animals, discuss the ethical considerations if these considerations are not adequately covered by the statement that you obtained approval from an ethics committee (see 'Materials' above).

The following example continues the short report quoted in the Introduction section:

The epidemiology of hay fever was studied in a national sample of 17 414 British children born during one week in March 1958 and followed up for 23 years (the National Child Development Study) I examined three outcomes:

(1) Self-reported 'hay fever during the last 12 months' at age 23 years.
(2) Parental report of 'hay fever or allergic rhinitis in the past 12 months' at age 11 years.
(3) Parental recall, elicited when the child was aged seven years, of 'eczema in the first year of life'.

The data on 16 perinatal, social, and environmental factors were cross-tabulated using the Statistical Analysis System. Multiple logistic regression models were fitted using the logistic regression program in the Biomedical Data Processing package. [Strachan 1989.]

Results

In the results section you are answering the question:

What did you find or see?

Write the section in such a way that it stands on its own, without the reader having to refer to other sections of the paper.

Decide on a logical order for the results and present them accordingly. The best order may be the order used for the protocols or procedures described in the methods section. Emphasize results that answer the question you are examining by putting those results at the beginning of a paragraph. Include results from controls as well as from experimental materials or subjects. Put secondary results after the more important results and then give any supporting information that is needed. Exclude results that are not relevant to your argument – but don't suppress valid results that appear to contradict your hypothesis. Suppressing such results is unethical. Instead, include those results and, if possible, explain why they are anomalous. Allow readers to assess your findings for themselves.

If you don't intend to discuss the findings in detail you may decide to combine the results and discussion sections. If you do this, state briefly what the findings mean as you present them.

Make it clear how the experiments or observations relate to your argument. As well as using tables and figures to present the data you collect, state the main results (the outcome of your analyses) in the text. That is, interpret the numbers for readers by writing, for example:

Y was lower in group Z than in the control group (6.0 [SE 0.5] versus 7.8 [SE 1.1] ng/ml).

rather than

In the control group Y amounted to 7.8 [SE 1.1] ng/ml. In group Z it was 6.0 [SE 0.5] ng/ml.

Don't, however, repeat in the text all the numbers that are presented in the tables and figures. Paper, printer's ink, and readers' time cost too much for this kind of duplication, and readers can see the data in the tables and figures for themselves. Similarly, don't repeat the table titles and figure legends in the text. That is, write 'Y was lower in group Z than in the control group (Fig. 1)', not 'The amount of Y found in group Z and in the control group is shown in Figure 1'. And don't use table titles or figure legends as topic sentences for paragraphs.

STATISTICS

Use parametric or non-parametric statistics as appropriate. Use well-known tests in preference to obscure ones, which make editors and referees suspect that your work may lack substance.

As well as explaining (in the methods section) how you transformed raw data into results you must (a) report the data in such a way that readers can assess the degree of experimental variation and (b) estimate the variability or precision of the findings. Use the standard deviation to show the variability among individuals, use the standard error of the mean to show the precision of the sample mean (Altman et al. 1983), and always state the number of measurements on which means are based.

Make it clear which measurements you are comparing with which. Summarize data by giving the mean and the standard deviation (SD) rather than by giving the mean and the standard error of the mean (SEM). If the data are from a skewed (asymmetrical) distribution, summarize them by giving the median and the interquartile range (the range between the 25th and 75th percentiles).

Some authorities recommend using '15.9 (SE 1.6)' in preference to '15.9±1.6' because the ± sign can cause confusion (Altman et al. 1983). Exact probability values such as $P = 0.34$ or $P = 0.02$ (given to no more than two significant figures) may be preferred to the conventional $P < 0.05$ (Altman et al. 1983) or to the simple statement that a finding was not significant.

Be careful to distinguish between your primary hypothesis and any exploratory analyses of the data that you undertook. Use two-sided tests whenever necessary (i.e. nearly always). Use confidence intervals to establish the degree of uncertainty in the findings (see Gardner & Altman 1989). For comparative studies the confidence interval should relate to the difference between groups.

Consult a standard statistics textbook for more detailed advice.

This example of a results section continues the short paper already quoted [tables not included here]:

Among the 16 factors studied, the most striking associations were between hay fever, number of children, and position in the family in childhood. At both 11 and 23 years of age hay fever was inversely related to the number of children in the household at age 11 (when most families were presumably complete). However, hay fever was strongly related to position in the family as well as to family size. When prevalence figures were adjusted by multiple logistic regression for other significant determinants of hay fever in this cohort, the associations with numbers of older and younger children in the household persisted. These trends in adjusted prevalence were independent of one another and each was statistically significant at the 1% level (see Tables 1 and 2 for P values), but the number of older children in the household appeared to be more important than family size ($\chi^2 = 11.6$, d.f. $= 1$, $P = 0.0007$ at age 11; $\chi^2 = 19.5$, d.f. $= 1$, $P = 0.00001$ at age 23). A further analysis of hay fever at age 23 by birth order and number of older children in the household suggested that the number of older children was the more influential variable.

Eczema in the first year of life was also independently related to the number of older children in the household (Table 3). There was no association between infant eczema and the number of children born later to the family. [Strachan 1989.]

Making a preliminary presentation

Presenting your methods and results at an informal meeting (see Ch. 12) will provide you with useful comments and criticisms. Alternative interpretations may be suggested, studies you have overlooked may be brought to your attention, and other oversights may be pointed out. This is why it is a bad idea to dispose of experimental material etc. before drafting the paper. If possible, make such a presentation before you draft the discussion section, then decide whether you need to do more work before completing the draft. A good progression is to present your work at (1) an institute meeting, (2) a national meeting, and (3) an international congress – after which most of the bugs should have been caught.

Discussion

In the discussion section you are answering the general question:

What do your findings mean?

The discussion is where you answer the specific question or questions you stated in the introduction. Show how your findings relate to existing knowledge. Explain what is new in your work and say why your results are important, without making extravagant claims for them. Indicate what the next steps might be. You must also discuss other results and hypotheses that are relevant to yours. Discuss, too, any possible errors or limitations in your methods and assumptions if you have not already dealt with these in the methods section. Distinguish between facts and speculation, and be cautious in extrapolating to other species or conditions. Referees dislike speculation and may recommend revision, so you may be wise to avoid it.

Concentrate on the main lines of your argument. Avoid the temptation to refer to every detail of your work again: repeating the results section in the discussion is a common fault in drafts.

A useful way to open the discussion is to use the end of the introduction as your starting point. If you restate the question posed in the introduction, make sure the question is still recognizably the same one. Then provide your answer and explain it as necessary before going on to discuss other results.

To distinguish between results and the conclusions you are drawing from them, use the past tense for results and the present tense for general statements and conclusions (see Ch. 8, 'General advice on style in science'). Introduce your conclusions by using a strong verb such as 'show' or 'indicate'. Identify speculation by using 'might' with the verb (Zeiger 1991). Note that statistics can never prove anything, only indicate that the opposite is unlikely and give a measure of how unlikely.

This short discussion completes the paper already used as an example:

Respiratory symptoms have various popular names but variation in the reporting of symptoms is unlikely to explain why first children in this study were more likely to be infected with hay fever than later children, a finding that was independent of paternal social status. Although total family size might affect recall of infant eczema by parents seven years later, the number of older children in the household is less likely to have specifically affected recall. Similar gradients in hay fever and eczema with increasing family size have been reported in five-year-old children in a British cohort born in 1970 (4).

The above observations do not support suggestions that viral infections, particularly of the respiratory tract, are important precipitants of the expression of atopy (5). The findings could, however, be explained if allergic diseases were prevented by infection in early childhood, transmitted by contact with older siblings in unhygienic households, or acquired prenatally from a mother infected by contact with older offspring. Later infection or reinfection by younger siblings might confer additional protection against hay fever.

Over the past century, declining family size, improvements in household amenities and higher standards of personal cleanliness have reduced the opportunity for cross-infection within families with young children. These factors may have led to the more widespread clinical expression of atopic disease, and could explain why this expression emerged earlier in the better-off families (1), as seems to have occurred for hay fever. [Strachan 1989.]

Title

Working titles are rarely suitable for the final paper. The best time to write the final title is either after you have drafted the main sections of the paper or at the revision stage.

The title answers the question

What is the paper about?

The working title was intended to help you in the early stages of preparing for publication; the final version must help readers. It should therefore be interesting, concise, informative – and accurate enough for use in indexing systems and bibliographic databases containing titles alone as well as in databases containing titles and abstracts without the full text. Many readers will discover your paper by seeing it listed in *Current Contents* or a similar 'secondary service'. They will then judge the paper's relevance on the title alone. A misleading or fanciful title will attract the wrong readers or none at all.

Most journals prefer short titles, typically 100 characters (including the spaces between the words) or 10–12 words. Readers find titles longer than this difficult to grasp when browsing through journals or scanning lists of titles on paper or screen. Start by checking whether your target journal limits titles even more drastically – perhaps to 80 characters to fill one line of a standard computer screen. Keep the length down by cutting out trivial words and phrases that contribute nothing to the information in the title ('the', 'of', 'on', 'Notes on', 'An approach to', 'A study of'). Don't try to get round the restriction on length by using abbreviations, apart from any that are accepted as names (DNA, for example). And don't string nouns and adjectives together to remove an 'of' or an 'on'; stacking 'modifiers' up in this way hides the meaning, as in this example:

Report of a 1984 visitors' exit interview study.

This title would be easier to understand and more informative if it read

Label comprehensibility assessed by interviews at British Museum exits in 1984.

Don't use trade names, jargon (see p. 96), or outmoded terms in the title. Be specific, not general or vague. 'Membrane permeability in insects', for example, would be better rewritten as:

Amino acid activation of ion channels in locust muscle

or as

Ion channels in locust muscle: activation by amino acids.

Try to start with a significant word or phrase. Say what you studied and, if journal practice permits, indicate the results, as in these two examples:

Pollen morphology of *Saxifraga nathorstii* in Berglund resembles that of *S. azoides* and *S. oppositifolia* in Sudland

High incidence of multiple myeloma in north-west Ireland, 1982–1986

Many journals, however, ban titles like these that make claims about the findings in the paper or that are in sentence form, especially if the sentence ends with a question mark. Some journals also object to 'hanging' titles: those with a main title and a subtitle ('Multiple myeloma in north-west Ireland: high incidence recorded in 1982–1986'). Most titles should therefore be of this kind:

Pollen morphology of *Saxifraga nathorstii* in Berglund and of *S. azoides* and *S. oppositifolia* in Sudland

Incidence of multiple myeloma in north-west Ireland, 1982–1986

If you are writing a series of papers on the same subject, try to give each paper a separate title rather than using the same general title and numbered subtitles. Successive papers may not all be accepted by the same journal, or part 3 may be ready for publication before part 2 struggles through the review process or – worse – is rejected. You can link the paper to others in the same series by mentioning the others in a footnote on the title page or by citing them in the introduction. If you must use a numbered subtitle, keep the same co-authors in the same order; a series of papers with different first authors produces problems for readers and librarians.

In addition to the main title the journal may ask for a short title, perhaps 45–60 characters in length, for use as a page headline (a 'running head') or sometimes as a footline. If your main title is longer than the stated limit for the short title, provide a shorter version for this purpose.

See also Chapter 9, 'Title page'.

Abstract

An abstract is defined as 'an abbreviated, accurate representation of the contents of a document, without added interpretation or criticism and without distinction as to who wrote the abstract.' It should be 'as informative as is permitted by the type and style of the document; that is, it should present as much as possible of the quantitative and/or qualitative information contained in the document' (ISO 214:1976). In the abstract it is even more important than in other parts of the paper to keep sentences short and simple, dealing with just one topic each and excluding irrelevant points.

Abstracts are usually described as informative or indicative (descriptive) or as a mixture of informative and indicative. Informative abstracts are best for papers describing original research. Indicative or informative–indicative abstracts contain general statements about the subjects covered in the document and are used for field reports, for long papers such as review articles, and for books or chapters in books. Structured abstracts covering specified topics may be required by some journals, as discussed below. Abstracts for meetings are also discussed below.

INFORMATIVE ABSTRACTS

An informative abstract answers, typically in 100–250 words, the questions

Why did you start?
What did you do, and how?
What did you find?
What do your findings mean?

If your paper is about a new method or apparatus the last two questions might be changed to

What are the advantages (of the method or apparatus)?
How well does it work?

These questions are of course the ones answered in the different sections of the text, but readers often have no access to the full text or no time to read it. The abstract must therefore be written so that it can stand on its own, without the text. As the title and abstract are always read together, however, don't waste words by repeating or paraphrasing the title in the abstract. Keep to 250 words or less for an article of 2000–5000 words, and to about 100 words for a short communication, depending on the journal's requirements.

If the reason for doing the study is not clear from the title or the rest of the abstract, state the purpose. If the type of document (report of original research, review article, case history, etc.) is not clear from the title or the rest of the abstract, mention what it is early in the abstract. Say what you studied and what methods you used. Give your main findings concisely and summarize your conclusions, as in this example based on a paper entitled 'Lateglacial and early Flandrian chronology of the Isle of Mull, Scotland' (Walker & Lowe 1982):

Pollen-stratigraphical records were obtained from limnic sediments and 15 radiocarbon dates were established in samples of bulked gyttja material from four sites on the Isle of Mull. These indicated that wastage of the Scottish ice sheet was followed by an improvement in climate on the island at around 13 000 yr BP, allowing vegetation to develop. A deterioration in climate at or before 10 700 BP then led to severe periglacial conditions. Another improvement in climate at around 10 200 yr BP caused the final wastage of the Loch Lomond stadial glaciers. The plant succession of the early Flandrian period began with an *Empetrum* phase, followed by *Juniperus* expansion between 9600 and 9500 BP, establishment of open birchwoods by about 9300 yr BP, and immigration of *Corylus* around 8800 yr BP. The radiocarbon dates form the basis for the first Lateglacial and early Flandrian chronology for the islands of the Scottish Inner Hebrides. The precise timing of events during the early Lateglacial Interstadial remains enigmatic.

Try to include in the abstract all the main information covered in the paper. Be as brief and as specific as possible, and write with non-specialists in mind. Emphasize the different points in proportion to the emphasis they receive in the body of the paper.

Do not under any circumstances refer in the abstract to information that is not in the paper.

Generally speaking, a short abstract should be written as a single paragraph. However, split a longer abstract into two or more paragraphs if this is clearer for readers and is allowed by the journal. Number the paragraphs or sentences only if the journal requests this. Write complete sentences that follow each other logically; don't use telegraphese. When possible, use active verbs, and use the past tense for what was found. 'Use the third person unless use of the first person will avoid cumbersome sentence constructions and lead to greater clarity' (ISO 214:1976).

To help computerized text searching, use significant words from the text in the abstract. Avoid unfamiliar terms, acronyms, abbreviations, or symbols; if you must use them, define them at first mention. Use generic names, not trade names, for chemicals and drugs, except when trade names are the most accurate way to describe such substances (see 'Materials' above). Identify living organisms by their Latin names.

Don't include tables, diagrams, equations, or structural formulae in an abstract unless it is intended for consideration by a conference organizing committee rather than as part of a journal article (see 'Conference abstracts' below). Avoid citing other work; if you must include a citation, for example to a paper that inspired your investigation, include a short form of the bibliographic details in the abstract itself – 'as A.B. Smith pointed out (J Geogr Info 1990;20:11–13)' – for the benefit of readers who see the abstract alone.

INDICATIVE ABSTRACTS

Indicative abstracts for long articles such as review articles give readers a general idea of the contents of the paper but little, if any, idea of specific methods or results. An indicative abstract mistakenly written for the short paper abstracted above might read something like this:

> Pollen samples were taken from four sites on the Isle of Mull and radiocarbon dates were established. Climatic variations occurring between 13 000 and 10 200 yr BP are discussed. The plant succession of the early Flandrian period is described.

STRUCTURED ABSTRACTS

Some clinical journals now ask for 'structured abstracts' for reports of clinical trials and maybe for other kinds of contributions too. Structured abstracts usually contain a maximum of 400 words and in clinical journals are divided into sections with the headings Objective, Design, Setting, Patients, Treatment, Results, Conclusion. This sort of abstract is written mostly as a series of points, although the Results and Conclusion sections should be in sentence form. Structured abstracts may evolve into a new kind of publication, with the main text available only in electronic form – or they may disappear altogether. If your target journal wants a structured abstract the instructions to authors will tell you what headings to use and how long the abstract should be. Examples of abstracts in the journal will demonstrate what is required.

CONFERENCE ABSTRACTS

A few conference organizers ask for structured abstracts but most abstracts for meetings should be written in the same way as a conventional informative abstract – except that you may be allowed to include a table or graph if you can fit it into the space available.

Conference abstracts often have to be typed on a form or blue-lined sheet of paper supplied by the meeting organizers. If your abstract is accepted for the meeting, the typed form you submit may be used as camera-ready copy for the printed abstract. You must therefore follow the organizers' instructions closely, check the typing carefully, and keep the abstract within the required limits of length and position on the page. To make sure that the abstract will keep within those limits, type or print out a draft on a copy of the form before you prepare the final version.

Key words

Key words or phrases intended as indexing and cataloguing entries are often printed at the end of an abstract, or sometimes after the title in the journal's contents lists. If the journal asks for key words, choose the most important and most specific terms you can find in your paper. The journal may ask for these terms to be chosen from, for example, the Medical Subject Headings (MeSH) used in *Index Medicus*, or from a list such as those published in *Biological Abstracts* and *Chemical Abstracts*. Words that appear in the title of the paper should not usually be included among the key words, but check what your target journal asks for. Put the necessary number of key words in the place required – usually on the title page or at the end of the abstract. To help the right readers to find your paper, keep the key words specific. Include more general terms if your work has interdisciplinary significance.

Summary of a paper

A summary is not the same as an abstract, although some journals call abstracts of the articles they publish 'summaries'. Nor is a summary the same as the conclusions. Strictly speaking, a summary restates the main findings and conclusions of a paper and is written for people who have already read that paper. An abstract is an abbreviated version of the paper written for people who may never read the complete version (see p. 70).

Include a summary only if the journal asks for one. If the journal asks for a summary in a language that is not your first language, follow the advice given in 'Translations' earlier in this chapter.

Acknowledgements

Acknowledge – briefly – any substantial help received from organizations or individuals, whether they provided grants, materials, technical assistance, or advice. (Note that some journals ask for funding bodies to be named on the title page instead of being included in the acknowledgements.) Acknowledge all those who went out of their way to help you or who did most of the day-to-day work, but not necessarily those who did no more than their routine laboratory or office work. If you need to acknowledge the head of the institute or department where the work was done, include him or her in the acknowledgements section. Don't thank miscellaneous friends or relations who did not contribute directly to the work you are reporting. Make sure that all those you thank are willing to be thanked and that they approve of the wording of your acknowledgement. If you are including previously published material in your paper it is sometimes appropriate to acknowledge the copyrightholders in this section, if they agree.

Appendix

If you have decided to include an appendix containing supporting information (see p. 13), prepare it now. If the appendix includes reference citations, the bibliographic information should usually be included in the main reference list.

List of contents

Journals that publish long papers may ask for a list of contents that will not necessarily be printed with the article (if you are writing a book a contents list will certainly be needed). A contents list is easy to draw up. Include the headings of the main and second-level sections, and perhaps the third-level sections too. For contributions submitted on paper, give the page numbers on which each section or subsection starts, for the benefit of the editor and referees. Type the headings of the main sections at the left of the page, with the page numbers at the right. Indicate the rank of second-level headings by indenting them a few spaces more than the main headings; indent third-level headings, if you include them, even further than second-level headings.

BURYING THE DRAFT

When you have finished drafting the main text of the paper, don't start revising immediately. Revision, which is a type of editing, demands that you view the manuscript with a fresh eye, to find any faults. So put the draft or disk out of sight for as long as you can afford, burying it deep in a desk drawer or filing cabinet. When enough time has gone by, dig the paper out and reread it, correct any obvious mistakes, and compile and type the reference list in the way described in Chapter 6. Type or retype either the whole draft or any pages of a paper copy that have become illegible. Unless a submission deadline is now too close, bury the draft again for a week or two before revising it in the ways described in Chapters 7 and 8.

SUMMARY

(1) Find a time and place for writing, and collect all the materials you'll need;   (2) Start with the easiest section and write as simply as you can, without worrying too much about grammar, nomenclature, or references, but insert headings to help yourself and the readers; (3) Write the introduction, materials and methods, and results sections;   (4) Make a preliminary presentation of your work;   (5) Write the discussion section;   (6) Revise the working title;   (7) Write an informative abstract;   (8) Provide key words, if required; (9) Write the acknowledgements section and make sure that those you thank approve of the wording;   (10) Prepare an appendix, if needed;   (11) Prepare a list of contents, if needed;   (12) Bury the draft for a while.

CHAPTER SIX

Storing, choosing and styling references

Building a bibliographic database   Choosing references for your
paper   Citing unpublished work   Citing published work
Styling citations and reference lists   Typing and cross-checking
citations and reference lists

The main problem with references is that neither publishers nor journal
editors have yet agreed on a uniform style for citing or listing them.
Happily, computer programs exist that can reduce the chore of making
references match the idiosyncratic requirements of publishers and
editors. This chapter describes how to store bibliographic information
with or without such a program, and how to prepare references for a
particular paper.

BUILDING A BIBLIOGRAPHIC DATABASE

During your career you will be submitting papers to journals using
different reference systems. When you record details of your reading you
should therefore include all the bibliographic details the different
systems require.

Keep your collection of bibliographic information in a computer
system, if possible. An ordinary word processor can be made to do a
useful job but a database program is better, while a specialized bibliographic program for managing references is best of all. Specialized
programs not only produce reference lists in the format of your choice,
but can also substitute names for numbers, or vice versa, and rearrange
citations in the text at the touch of a few keys.

If you use a non-specialized database program for your bibliographical
collection, set it up with separate fields for each part of a reference.
Include fields for a reference number for each reference, and for authors'
names, date of publication, article title, journal title in full, volume

number, issue number, first and last page numbers, secondary source (see p. 78), key words, and your own summary or comments. For references to books add fields for editors, series editors, place of publication, publisher, series title, and anything else you know is regularly included in published reference lists in your branch of science.

If you are not using a computer, type the bibliographic details on cards and file the cards alphabetically under the first author's name. Or make two cards for each reference and file the second by subject, to make retrieval easier if you forget the authors' names.

If you don't possess a specialized bibliographic program a useful way to record references is shown in examples (a)–(f) below (based on ELSE-Ciba Foundation Workshop recommendations, 1978 and ISO 690:1987; see also BS 1629:1989). Style (a) is for journal articles:

(a) Lovelock J E, Whitfield M. 1982. Life span of the biosphere. Nature (London) 296:561–563.

Styles (b), (c) and (d) are useful for recording references to books, chapters in books, and chapters in books in a series:

(b) Howard J. 1982. Darwin. Oxford: Oxford University Press. 101 pp.

(c) Heslop-Harrison J. 1983. The scientific information system in the United Kingdom. In: Manten A A, Timman T, eds. Information policy and scientific research. Amsterdam: Elsevier, pp. 113–118.

(d) Fahrbach S E, Truman J W. 1987. Mechanisms for programmed cell death in the nervous system of a moth. In: Bock G, O'Connor M, eds. Selective neuronal death. Chichester: Wiley, pp. 65–76 (Ciba Foundation Symposium 126).

Record titles in the language of publication. If your typewriter/printer can't cope with the original language, put the transliterated title between brackets. Use brackets for a translated title if you give this in addition to the original title, and put additional information in parentheses:

(e) Patate J-G. 1990. Zut alors! [Good gracious me!]. Revue française des idiotismes 123:456–478. (In French with English abstract.)

Record the names of authors in the way they are printed on the title page of the article, chapter or book, but put family names before initials, as in the examples above. Write one given name (forename) in full if your target journals are likely to want this information – a few do, though in most journals initials alone are sufficient.

Copy the titles of books from the title pages, not the covers. If several cities of publication are listed on the title page include just the first city and use a short form of the publisher's name (Wiley, or John Wiley, not John Wiley & Sons Ltd) unless there are two or more publishers with similar names; when in doubt, use the full name. If the book is part of a series, include this information too, as in example (d) above.

Give the year of publication as printed on the journal cover or on the copyright page of a book, not the date a paper was submitted to a journal or the date a conference was held. If the date of issue of the publication is later than the cover date, cite the actual date of publication, for reasons of priority, if you can establish what it was (some journals state the publication date in the next issue).

Include the issue number of a journal if each issue rather than each volume starts with page 1, and give the date of publication – day and week or month – if there is no volume number:

(f) Smith M J, Jones J P, Brown M H. 1986. How to grow old happily. Journal of the Royal Society of Ageing 123(2):145–167. [Or 'p 145–167 (17 March).']

For examples of other kinds of references see Tables 6.2, 6.3 and Appendix 1.

Ideally you will read all the original papers yourself but if you want to keep details of an article, an abstract, or an extract that you haven't read but have seen referred to in another publication, name the secondary source as well as the primary reference. That is, add '[cited by X 1986]' or '[abstract in Y 1987]' to your record of the primary reference and include details of the 'X 1986' or 'Y 1987' reference in your collection of references.

If you are recording details of unpublished material, include as many bibliographic details as are available and indicate that the material is not published. For information learned at a meeting, for example, give the date, place, and full title of the meeting, as well as the speaker's name. (See 'Citing unpublished work', below.)

If you haven't got easy access to a good library or to a printed or online bibliographic database, keep reprints or photocopies of the most important papers you read. If no bibliographic information is printed on the original paper, write the details on the first page of your copy as soon as you make it or obtain it, and check all the details carefully against the original publication before returning it to the library or other owner. If you obtain information from an online bibliographic or full-text database, make sure you transcribe the necessary information correctly or download it properly. Check the details against the original publication too, when you get the opportunity – neither databases nor other people's reference lists are completely trustworthy.

CHOOSING REFERENCES FOR YOUR PAPER

When the first draft of your paper is complete, including the tables and figures, collect the necessary bibliographic information and start preparing the reference list. First check whether your target publication limits the number of references per paper or the number allowed in support of any one statement. You must include a citation whenever you mention previous work by others or by yourself, but be selective – don't include everything ever written on the subject. Cite one or two review articles covering the general background, for the sake of readers who want to investigate that background, but make sure that most of your reference list consists of reports of original research.

Reread all the references you have cited and decide whether you should have cited them and whether others should be added now. Make sure that the work you cite is indeed relevant to the points you are making and that you understood it properly at your first reading. You won't impress the editor, the referees, or the readers if you cite material that is only loosely related to what you are talking about or if it is clear that you misunderstood it. Readers will also think better of you if you cite other people's work (preferably theirs), not just your own.

Don't cite material you haven't read yourself unless it is impossible to obtain the original document; if you must cite such a publication, cite it in the text as '(Teodorescu 1984, cited by X 1986)'. Give both references in the reference list and add '[cited by X 1986]' at the end of the unread reference.

CITING UNPUBLISHED WORK (TABLE 6.1)

The instructions to authors may tell you how to handle unpublished references. Many journals do not allow unpublished work to be included in reference lists. Instead, relevant details should be included in the text, for example as

A.B. Jones, unpublished lecture on 'Junk in space', University of Birmingham, 21 July 1999.

Readers then know immediately whether statements are supported by refereed work to which they can turn if they want.

Table 6.1 Citing unpublished material.

Source	Permission needed to cite	Cite in text	Include in reference list
Public archives	no	yes	yes
Private collection	yes	yes	no*
Confidential or in preparation	yes	yes	no
Telephone conversations†	yes	yes	no
Technical reports, etc.			
Available on request	yes	yes	yes
Not available on request	yes	yes	no
Theses	no	yes	yes

* May be included in reference list if the collection is open to the public.
† Send a copy of the wording to the other person and obtain his or her written permission. Do not cite if permission is not granted.

Always state that the material is unpublished. Note that many journals treat conference abstracts as unpublished, because these are not always subject to a selection procedure and are not always followed up by full papers.

If you cite unpublished material from public archives you can legitimately include a reference in the reference list (if the journal has no objections), because the material is publicly available. If you want to cite unpublished material from a private collection, obtain permission to do so, and cite it in the text only, not in the reference list, unless the private collection is open to the public. If you cite documents that are not publicly available, whether because they are confidential or because they are still being prepared for publication, obtain permission to refer to them and include them in the text only. Describe them there as 'X.Y. Smith, personal communication' or as 'A.Z. Brown, unpublished work, 1991', with other details, if available, and preferably with a date. Don't use 'in preparation', which editors and referees disapprove of unless you can provide a copy of the manuscript; and don't use 'private communication', which sounds as if you are giving away a secret. Don't cite telephone or other conversations as personal communications unless you have sent a copy of the wording to the other person and obtained his or her written permission to refer to the conversation.

Avoid references to technical reports or similar documents of limited circulation (known as the 'grey literature') unless you have obtained permission to refer to these documents. If this literature is available on request, include a reference number and the full address of the source with the other details in your reference list. If a document is not available on request, cite it in the text only, as for unpublished work.

Doctoral and other theses can be included in the reference list because they are publicly available, though not always easy to obtain. If the contents have been published as journal articles, however, cite the articles rather than the theses.

CITING PUBLISHED WORK

Many scientific publications use the name-and-year (Harvard) system in which authors' names and the date of publication are cited in the text in one of these ways:

Black and White (1991) have suggested that . . .

As already reported (Black and White 1991) . . .

In this system the reference list is arranged alphabetically (see examples in Table 6.2).

Table 6.2 Name-and-year (Harvard) style for references*.

Type of reference	Example†
Single-author paper in journal	Adam A. 2000. Preventing rot in Mediterranean apple trees. Journal of Tree Studies [or J Tree Stud] 99:21-29.
Single-author book	Adam A. 1999. Treatment of fungus in apple trees. New York: Wiley. 300 pp.
One-page paper in edited book	Back V F. 1977. Seeds of *Stellaria media* (L.) Vill. In: Smith J, ed. Anatomy of Caryophyllaceae. Berlin: Springer, p. 250.
Paper in edited book in a series	Bull F. 1975. Jumping over the moon. In: Jones M, Lloyd P, eds. Space travel. Houston: Galaxy Press, vol. 3, pp. 1-24 (Soc Space Sci Symp 21).
Paper in journal, two authors	Lovelock J E, Whitfield M. 1982. Life span of the biosphere. Nature (London) [or (Lond)] 296:561-563.
Multi-author technical report	Lovelock J E, Willis A, Ziegler Y. 1990. Environmental research in Wisconsin. Chicago: Metallurgical Processing Corporation. METPRO/CB/TR--90/256. 150 pp.

* An alphabetic-numeric reference list would be similar to the list of examples in this table but would have a number placed before each reference.
† Punctuation, typography, and other details may vary from journal to journal.

Many other publications use the sequential-numeric system in which numbers are used for citations in the text:

The following theory was suggested recently[1] . . .

or

As Smith[1] has suggested . . .

In this system references are numbered and arranged in the reference list in the order in which they are first referred to in the text (see Table 6.3), although a few journals ask for new numbers to be assigned to the second and each further citation of the same reference. The 'Vancouver style' used by many biomedical journals is a sequential-numeric system (see Appendix 1).

Some publications favour the hybrid alphabetic-numeric system in which the reference list is arranged alphabetically and then numbered; the numbers assigned in the list are then used for citations in the text.

Other variations include reference lists in which the references are given in date order, with the year of publication appearing first in each reference entry. A few publications, especially in chemistry, put

Table 6.3 Sequential-numeric style for references.

Type of reference	Example*
Single-author paper in journal	1. Adam A. Preventing rot in Mediterranean apple trees. Journal of Tree Studies [or J Tree Stud] 2000;99:21-29.
Single-author book	2. Adam A. Treatment of fungus in apple trees. New York: Wiley, 1999. 300 pp.
One-page paper in edited book	3. Back V F. Seeds of *Stellaria media* (L.) Vill. In: Smith J, ed. Anatomy of Caryophyllaceae. Berlin: Springer, 1977, p. 250.
Paper in edited book in a series	4. Bull F. Jumping over the moon. In: Jones M, Lloyd P, eds. Space travel. Houston: Galaxy Press, 1975, vol. 3, pp. 1-24 (Soc Space Sci Symp 21).
Paper in journal, two authors	5. Lovelock J E, Whitfield M. Life span of the biosphere. Nature (London) [or (Lond)] 1982;296:561-563.
Multi-author technical	6. Lovelock J E, Willis A, Ziegler Y. Environmental research in Wisconsin. report Chicago: Metallurgical Processing Corporation, 1990. METPRO/CB/TR--90/256. 150 pp.

* Punctuation, typography and other details may vary from journal to journal. Further examples are included in Appendix 1.

references in footnotes or side notes on the pages where they are cited, with or without a reference list at the end of the article.

Whichever reference system your target journal uses, it is usually more practical to use names and years than numbers in the early drafts of your paper (see Ch. 5, 'References').

STYLING CITATIONS AND THE
REFERENCE LIST

Even publications that use the same reference system have very different ways of dealing with punctuation, typographic appearance, and the sequence of the different parts of each reference. In general, follow the instructions to authors or copy the style of reference lists printed in the journal. If you are not using a bibliographic program and if parts of the references are to be printed in italics, underline those parts if you are asked to do so. Put wavy lines below any parts to be printed in bold – again only if you are asked to do so.

If the journal uses the Vancouver style, follow the examples of this style exactly as they are printed in the journal's instructions to authors rather than copying the style used in printed reference lists in the journal. Many journals that accept the Vancouver style in fact publish their own variations on it (Porcher 1986) – but there is no need for you to work out what those variations are.

In the reference list put family names before initials, unless it is journal style to invert only the first author's name and initials. Be careful with oriental names: Chang Guang-sheng should be listed under C, not G, but sometimes such names have been westernized, becoming G.S. Chang (see CBE Style Manual Committee 1983 for more detailed advice).

In alphabetical lists, follow the journal's usual system for alphabetizing references that have the same first author. If no system is obvious, arrange such references alphabetically by the first author's name and then in this order: (1) alphabetically by the second author's name if there are only two authors; (2) in date order (earliest date first) if there are three or more authors and if 'X et al.' is used for all such references in the text. Always follow the principle that authors' names should be given in the same form in the reference list as they are written in the text.

If the instructions ask for journal titles to be abbreviated, note whether and where full stops (periods) and capital letters should appear and which abbreviation system is used. The most widely-used abbreviation system for title words complies with an international standard (ISO 4:1984; and see BS 4148:1975). This 'International List' system (ISDS 1975) is used in the annual list of journals indexed by *Index Medicus*, in the

Chemical Abstracts Service Source Index (CASSI), and in *Serial Sources for the BIOSIS Previews Database*, amongst other places (and see Huth 1987 for a useful list of abbreviations of words used in journal titles). If no list of approved abbreviations is available, write all journal names in full: it is easier for editors or copy editors to delete text than to add it (and it is easier for them to add punctuation than to delete it). Do not use the five- or six-letter names included in the CASSI and BIOSIS lists to represent journal titles; these names, known as CODENS, are machine-readable but can't be decoded easily by most human readers.

Citations in the text: name-and-year system

If there are three or more authors, or four or more authors, some publications that use the name-and-year system print all the names the first time such references are cited (Black, White & Green 1988) and then print the first author's name and 'et al.' if the same reference is cited again (Black et al. 1988). Other publications use 'Black et al.' every time a reference with three (or four) or more authors is cited.

'Et al.' is short for 'et alia' ('and others') and should never be used to represent just one name: 'Black et al.' is acceptable for 'Black, White & Green' but not for 'Black & White'. When 'et al.' is used in the text the authors' names, or as many of them as the journal prints, must still be included in the reference list.

If no author or editor is named, choose an appropriate word or words from the title of the organization or group that produced the document, such as 'Meteorological Office (1987)', and use the same name in the reference list. If you cannot find a suitable name, invent a term such as 'Birmingham Survey' to link a citation in the text to the reference in the reference list. Try to avoid using 'Anon.' or 'Anonymous'.

When you cite several references together, put the most important one first. If you think they are all of equal importance, put them in chronological order, with the earliest date first.

When you refer to a specific page of a long article or book chapter the journal may ask for the page number to go in the text: 'This was reported by Smith (1974, p. 15)'. In the reference list include the first and last pages of the article or chapter if that is journal style, but not the particular page(s) you referred to in the text.

If you refer to two or more papers published in the same year by the same author, or to several papers by the same first author with co-authors who are identified only as 'et al.', add the letters 'a', 'b', 'c' ... to the year (Black 1986a, b, White et al. 1987a, c) unless your target publication prefers another style.

Citations in the text: numeric systems

Write numbers for citations in the text in the journal's usual style – above or on the line of type, with or without parentheses or brackets. If you cite a particular page in a book or a long article, include the page number in the citation by writing, for example, '(Ref. 12, p. 15')'.

TYPING AND CROSS-CHECKING CITATIONS IN THE TEXT AND REFERENCES IN THE LIST

When you have compiled your reference list, type it (or print it out from your bibliographic program) in journal style, double-spaced – that is, with about 30 lines to an A4 or 8.5 by 11-inch page. Don't type 'idem' or 'ibid.' or put dashes or ditto marks in place of authors' names when the same name(s) appear in successive references – unless this is journal style: references may get moved or removed and the 'idem' or 'ibid.' reference will then be assigned to the wrong author.

Check the typing carefully, especially names, technical terms, foreign words, dates, volume numbers and page numbers, and any other material a copy editor cannot check easily. In other words, check EVERYTHING carefully . . . Then make sure that every citation in the text has a corresponding entry in the list and that every entry in the list is cited in the text. Remove redundant references from the list and add any references that are missing.

With the name-and-year system make sure, for each reference, that the names are spelt the same way in the text and in the list and that the date (year) is the same in both places. If you find discrepancies, check the bibliographic details again. If possible, look at the original publication when you make this check, or at least use a reliable secondary source such as *Index Medicus, Current Contents, Biological Abstracts* or *Chemical Abstracts*. If you don't possess a bibliographic program that can alphabetize the names and arrange references in the required style, make sure (a) that the reference list really is alphabetically arranged in the way required by the journal, (b) that all the required information elements are included for each reference, and (c) that the journal's preferred style for the sequence of elements and for punctuation and typography has been followed.

With the sequential-numeric system check that the numbers in the text run in sequence, that each number has a matching reference in the list, and that the reference is the one you intended to refer to. As with the name-and-year system, make sure that all the necessary bibliographic elements have been included and that you have followed the publication's instructions exactly.

With the alphabetic-numeric system, check that each reference number in the text corresponds to the correct reference in the list and, again, that all the necessary bibliographic information has been given in the required form.

If you think all these details are tiresome, you're right. But publication of a paper can be delayed for weeks or months if a reference or even a page number is missing. Worse, referees are suspicious of papers with carelessly presented references because this implies a similarly slipshod approach to other details such as the statistical analysis. Making sure that references are correct and complete is in your own interests, as well as being a courtesy to readers.

SUMMARY

(1) Collect and store full details of your reading;   (2) Select references for your paper;   (3) With certain exceptions, cite unpublished work in the text only;   (4) Cite published work in the text in the style of your target journal;   (5) Arrange and type the reference list in the style of your target journal;   (6) Cross-check citations in the text and references in the reference list.

CHAPTER SEVEN

Revising the first draft: content and structure

Logic and order   Tables and figures   Citations and quotations
Nomenclature, abbreviations, and footnotes   Retyping the draft
Checking length

When you dig up your draft after its burial period (Ch. 5, 'Burying the draft') your next job is to revise it. Concentrate first on the structure, as described in this chapter. All the parts, paragraphs, and sentences must be in the right order before you revise the style (Ch. 8) – but make a list of any stylistic problems you notice while examining the structure. Make sure that all the essential points you want to make have been included and any superfluous ones removed. Check that the argument runs logically from beginning to end – from hypothesis to conclusions. Ask yourself again whether you had something worth writing about. Then make sure that you have indeed said what you meant to say.

LOGIC AND ORDER

Examine the draft first for logical necessity, order, accuracy, consistency, and truth. Everything you say should contribute in some way to your argument and no steps in the argument should have been left out. But during the drafting process you may have wandered away from your main argument, introduced unnecessary material, left out essential evidence, or discussed points in the wrong order. Check these matters now, especially whether some passages would be clearer if you moved them to another place in the text.

Make sure that the headings relate properly to one another and to the text they describe. Should any headings be deleted or new headings added? Is each heading appropriately ranked? Is each one identified – by a marginal note or by its typographical appearance – as a first-order, second-order, or third-order heading?

Don't jump backwards and forwards between several different ideas . . .

Another way of examining the structure of the paper for logical flow is to see how long the paragraphs are and how the ideas are distributed among them. In principle, each paragraph should cover a single topic or message and be a 'unit of thought'. In practice, paragraphs tend to describe various characteristics of the topic under discussion or move from one argument to another – from premise to conclusions, from specific to general (or vice versa). In such cases it is important to deal fully and finally with one characteristic at a time. Don't jump backwards and forwards between several different ideas in either the same sentence or the same paragraph.

Readers need to rest their eyes on white space from time to time, so look for suitable places to break paragraphs that are longer than 125 words or so (about half a typewritten page, double-spaced). On the other hand, if many of your paragraphs consist of only one or two sentences, combine most of them – provided that each new paragraph deals with one main concept only.

TABLES AND FIGURES

When you are happy with the structure of the text, look at the draft tables and figures again. If any of them are redundant or less relevant than you thought at first, remove them. If some of the remaining tables or figures can be combined for greater effect, or if some of them need to be simplified, make the necessary changes, including appropriate changes to table titles, footnotes, and legends. Then prepare the final versions in the ways described in Chapters 3 and 4. Reconsidering your results before preparing the final versions will give you a new view of the text – which may afterwards need further changes.

ACCURACY AND CURRENCY OF CITATIONS
AND QUOTATIONS

Reread the articles and other publications you cite in the draft. It may be a long time since you read some of them and if you rely on memory alone you may misquote or misrepresent the work described – even your own work. Check the methods sections of the articles particularly carefully: the methods you are using now may have wandered a long way from the methods you thought you were following in every detail. Correct the draft text accordingly.

Check recent issues of the main journals in your discipline for new papers on the subject. Notes added at proof stage are expensive and can give readers a bad impression of the way you work. Referees, of course, always seem to have read the latest papers.

If you quote directly from your own or someone else's work, quote the passage in context to avoid distorting its meaning. Keep to the letter of the original as well as to its spirit. Reproduce quotations exactly as printed in the original, including any mistakes; insert '[sic]' after any word or phrase you think was misspelt or misused in the original. If you add anything to the quoted passage place the added words or characters between square brackets. Use three spaced stops (ellipsis points: . . .) to show where words have been left out in the middle of a quotation.

If a quoted passage is not in English, translate it; include the original version only if you feel it is essential to do so. Get your translation checked by a native speaker and insert ['My translation']. Make sure that the entry in the reference list uses or states the language in which the paper was written.

Obtain permission from the copyrightholder for quotations of more than 100 words or 5% of the original article, whichever is less (see Ch. 2, 'Coping with copyright').

NOMENCLATURE, ABBREVIATIONS, AND FOOTNOTES

Check that the nomenclature you use is up to date and approved by the appropriate authorities in your discipline. A few of those authorities are listed here (Table 7.1); your target journal may name others in its instructions to authors. (See also Ch. 8, 'Typography').

Use Système International (SI) units and the standard abbreviations for them throughout.

Check that you haven't used too many abbreviations, even those approved by your target journal. You can legitimately use abbreviations to replace lengthy terms that appear more than about ten times in a

Table 7.1 Nomenclature and terminology: some general sources*.

Agronomy
American Society of Agronomy et al. 1984. Handbook and style manual. Madison, WI: American Society of Agronomy, Crop Science Society of America, Soil Science Society of America.

Anatomy
International Anatomical Nomenclature Committee/Subcommittees. 1989. Nomina Anatomica, 6th ed; Nomina Histologica, 3rd ed; Nomina Embryologica, 3rd ed. Edinburgh: Churchill Livingstone.

Bacteriology
Krieg N R, Holt J G, eds. 1984 &1986. Bergey's manual of systematic bacteriology, 2 vols. Baltimore, MD: Williams & Wilkins.

Skerman V B D, McGowan V, Sneath P H A, eds. 1980. Approved lists of bacterial names. International Journal of Systematic Bacteriology 30:225–240.

Biochemistry
IUB [International Union of Biochemistry]. 1978. Biochemical nomenclature and related documents. London: Biochemical Society. [For further IUPAC-IUB Joint Commission on Biochemical Nomenclature recommendations on nomenclature see issues of the *European Journal of Biochemistry*.]

Webb E C, ed. 1984. Enzyme nomenclature: recommendations (1984) of the Nomenclature Committee of the International Union of Biochemistry. Orlando, FL: Academic Press.

Botany
Mabberley D J. 1987. The plant-book: a portable dictionary of the higher plants. Cambridge: Cambridge University Press.

Chemistry and pharmacology
Dodd J S, ed. 1986. The ACS style guide: a manual for authors and editors. Washington, DC: American Chemical Society.

IUPAC [International Union of Pure and Applied Chemistry]. 1971. Nomenclature of inorganic chemistry, 2nd ed. Oxford: Pergamon Press. [Reprinted 1981.]

IUPAC. 1979. Nomenclature of organic chemistry. Oxford: Pergamon Press.

Marler E E J, compiler. 1985. Pharmacological and chemical synonyms: a collection of names of drugs, pesticides and other compounds drawn from the medical literature of the world, 8th ed. Amsterdam: Elsevier.

The Merck index: an encyclopedia of chemicals, drugs, and biologicals, 11th ed. (ed. Budavari S et al.) 1989. Rahway, NJ: Merck.

Reynolds J E F, ed. 1989. Martindale: the extra pharmacopoeia, 29th ed. London: Pharmaceutical Press.

Earth sciences (general)
Cochran W, Fenner P, Hill M, eds. 1984. Geowriting: a guide to writing, editing,

Table 7.1   continued.

and printing in earth science, 4th ed. Alexandria, VA: American Geological Institute.

Geology
Bates R L, Jackson J A. 1987. Glossary of geology, 3rd ed. Alexandria, VA: American Geological Institute.

Dutro J T, Jr, Dietrich R V, Foose R M compilers. 1989. AGI data sheets for geology in the field, laboratory, and office. Alexandria, VA: American Geological Institute.

U.S. Geological Survey. 1978. Suggestions to authors of the reports of the United States Geological Survey, 6th ed. Washington, DC: U.S. Government Printing Office.

Immunology
Rosen F S, Steiner L, Unanue E R. 1989. Macmillan dictionary of immunology. London: Macmillan.

Life sciences (general)
CBE Style Manual Committee. 1983. CBE style manual: a guide for authors, editors, and publishers in the biological sciences, 5th ed. Bethesda, MD: Council of Biology Editors.

Huth E J. 1987. Medical style and format: an international manual for authors, editors, and publishers. Philadelphia: ISI Press [now available from Williams & Wilkins, Baltimore & London].

Mathematics
ISO 31-11:1978. Mathematical signs and symbols for use in the physical sciences and technology. Geneva: International Organization for Standardization.

Swanson E, ed. 1979. Mathematics into type. Providence, RI: American Mathematical Society.

Medicine
Iverson C, Dan B, Glitman P et al. 1989. American Medical Association manual of style, 8th ed. Baltimore: Williams & Wilkins.

Microbiology
American Society for Microbiology. 1985. ASM style manual for journals and books. Washington, DC: American Society for Microbiology.

Physics
American Institute of Physics. 1978. Style manual. New York: American Institute of Physics.

Institute of Physics. 1983. Notes for authors. London: Institute of Physics. 36 pp.

Psychology
American Psychological Association. 1983. Publication manual of the American Psychological Association, 3rd ed. Washington, DC: American Psychological Association.

Table 7.1   continued.

Virology
Hull R, Brown F, Payne C. 1989. Directory & dictionary of animal, bacterial and plant viruses. London: Macmillan.

---

* Numerous sources of detailed information on nomenclature, mainly in the life sciences, are listed in the *CBE Style Manual* (CBE Style Manual Committee 1983, p. 155–243) and in *Medical Style & Format* (Huth 1987, p. 126–138 and p. 211–259).

ten-page manuscript, or that appear several times in quick succession, but don't use more than four or five such abbreviations in a single paper. And don't make sentences indigestible by using too many abbreviations in a short space:

> MPTP is converted by MAO-B to MPP, which reaches SNpc nerve cells via DA uptake systems

may be perfectly intelligible to expert colleagues but will be unacceptable to others. Match your style to the journal's readership. Using the full terms rather than too many abbreviations will help you to keep the meaning of sentences clear in your own mind.

Some abbreviations need not be spelt out. Chemical and mathematical symbols, being internationally understandable and unambiguous, are nearly always acceptable, although they are not always recommended for use in the body of the text. Make sure that any other abbreviations you use are approved by the appropriate authority in your discipline. Define these and other essential abbreviations at their first appearance, or in a footnote at the beginning of the paper, or in both places, according to the journal's requirements.

Once you have defined an abbreviation, use it whenever you need it – don't switch back to using the full term unless many pages have elapsed since its previous appearance, when you may remind the reader, once, what the abbreviation means. If you use – and define – an abbreviation in the title of a paper (although this is not recommended: see Ch. 5, 'Title'), redefine it in the text. Do the same for abbreviations used (and defined) in the abstract.

Avoid using footnotes. Even if you are writing for a journal that allows footnotes, use them only when you want to include subsidiary information that would seriously interrupt your argument if it was in the text. If the footnotes you propose to use are short enough (a sentence or two), put them between parentheses in the text instead of as footnotes.

If you are using endnotes, remove any that are not essential. (Endnotes are notes, sometimes including or consisting of references, that are printed at the end of a paper or chapter rather than on the page where they are referred to.)

TYPE OR RETYPE THE DRAFT

At this stage you may need to retype the manuscript or correct the draft on your word processor, to have a clean copy to work on when you revise the style. If you type or correct the draft yourself you are more likely to notice where improvements can be made than if you simply read through the draft to see whether the corrections will make sense to a typist. If you have made a lot of changes on a word-processor printout it will probably be quicker to type the whole article again than to correct it line by line on the screen.

Use good paper and a well-inked ribbon (where necessary) for printing the second draft, with double or even triple spacing and wide margins all round to give yourself plenty of room for making more changes.

CHECK THE LENGTH OF THE PAPER

Before revising the second draft for style, check its length against the journal's requirements – sometimes given as a page limit but often stated as a certain number of words, tables, and figures; sometimes the number of references is limited too. If your word processor doesn't count the number of words for you, there is no need to count every word in the text. Instead, count the number of characters in a typical line, divide that number by six (five characters is the average length of words in English, plus one for the space between words), and multiply the result by the number of lines on a typical page and by the number of pages in the text:

$$(\text{Characters/line}) \times 6 \times \text{no. of lines} \times \text{no. of pages}$$

You may then need to condense the paper further before it is acceptable to the editor. Don't plan to use tiny print, narrow spacing, inaccurate page numbering, or other trickery to make your manuscript seem to meet a page limit – editors recognize such devices very easily.

SUMMARY

(1) Examine the text for logical necessity, order, accuracy, consistency, and truth;   (2) Check that the draft tables and figures are

necessary and relevant, and prepare the final versions;   (3) Check the accuracy of citations and quotations;   (4) Check whether new papers on the subject have appeared;   (5) Check that nomenclature is correct and up to date;   (6) Reduce the number of abbreviations and footnotes/endnotes;   (7) Retype the draft in double or triple spacing on good quality paper;   (8) Check the length of the draft.

CHAPTER EIGHT

Revising the second draft: style

General advice on style in science   Problems of grammar and style
Basics of technical style   Obtaining comments on the second draft

Editors hope to receive adequately written papers from native English-speakers but they don't expect great writing from them. Nor should editors demand perfect English from ESL authors (those for whom English is a second language – or even a third or fourth). They do, however, want all authors to write as simply and clearly as possible and avoid major errors in grammar.

This book doesn't pretend to be a textbook of English grammar. It can only suggest ways of avoiding the mistakes that authors of all nationalities tend to make. If you need advice on the finer points of English usage see Strunk & White (1978) and Fowler (1965) or Follett (1974) or Gowers (1986). Consult the *CBE style manual* (CBE Style Manual Committee 1983) or *Scientific writing for graduate students* (Woodford 1968) for further advice on scientific prose. Benjamin's *Elementary primer of English grammar* (1989) or Gordon's *The transitive vampire* (1984) will be helpful if you need a modern grammar textbook, while Roberts' (1987) *Plain English: a user's guide* is a useful paperback covering grammar, vocabulary, and style, among other matters.

One reason for trying to avoid grammatical errors is that clarity and correctness encourage people to read your papers and help them to understand what they are reading. In general, you should simplify and shorten the text wherever possible. Decide what your goal is – what message you want to get across – and aim to reach that goal by the most direct route you can find. Editors, referees, and readers will all be grateful if you write simply, concisely, and correctly. Clearly expressed and logically ordered ideas are, however, more important than perfect grammatical form, which someone else (perhaps even the editor) can help you to achieve if you make your message clear.

. . . the mistakes that authors of all nationalities tend to make.

GENERAL ADVICE ON STYLE IN SCIENCE

(1)   *Use the first person* ('I' or 'we') for describing what you did – but don't overuse it, and don't use it if the journal or your supervisor has banned it.

(2)   *Use the active voice* ('X crossed the membrane') in preference to the passive voice ('The membrane was crossed by X'). Over-indulgence in the passive is the main cause of dullness in scientific writing. But don't go to the extreme of removing the passive completely: use it when readers don't need to know who or what performed the action, as in 'The animals were fed at four-hour intervals' (and see 'Verbs', section e, below).

(3)   *Use the past tense* for observations, completed actions, and specific conclusions ('The infusion caused local irritation').

(4)   *Use the present tense* for generalizations and statements of general validity ('Most regions where this problem arises belong to category X').

(5)   *Avoid 'gobbledegook jargon'* – the 'pompous use of long words, circumlocution, and other linguistic flatulence' (Howard 1984, p. 44):

> Although solitary under normal prevailing circumstances, racoons may congregate simultaneously in certain situations of artificially enhanced nutrient resource availability

is a fine example of gobbledegook jargon that cries out to be simplified ('Racoons live alone but come together when bait is provided'). The other kind of jargon – the specialist vocabulary of different groups or disciplines – is a permissible or even essential shorthand. Don't let this technical jargon turn into gobbledegook, which it does all too easily.

(6)   If you are an ESL author, *don't apply the same principles of style when you write in English* as are used in your language. If you need to, borrow technical phrases – but never whole sentences – from articles by English or American scientists in well-edited journals.

If you need scientific models for your writing style, read – for example – Lewis Thomas, Peter Medawar, Richard Dawkins, or Stephen Hawking. Or see *New Scientist* and other science magazines for good semi-popular writing that may help you to keep your style down to earth.

PROBLEMS OF GRAMMAR AND STYLE

Peter Woodford (1968), in providing guidelines for solving problems of style, suggested these four principles:

(1) Be simple and concise   (2) Make sure of the meaning of every word   (3) Use verbs instead of abstract nouns   (4) Break up noun clusters and 'stacked modifiers' (that is, strings of adjectives and nouns, with no clues about which modifies which)

Woodford's principles should help you to avoid most of the weaknesses that make scientific writing difficult to read. The four principles summarize the rest of this chapter, which covers the points listed in Table 8.1. Table 8.2 explains grammatical terms that are not defined in the text.

If you use a computer to prepare your manuscript, a style-checking program such as Grammatik IV or RightWriter may help you to avoid some obvious faults of the kinds discussed below. A style-checker won't find all the mistakes, though, and may even produce extra problems for you. If you use such a program, do so with care. Note that many style programs merely assess style without pointing out grammatical errors or suggesting how you might improve your writing or correct faults.

Verbs

(a) DO SUBJECTS AND THEIR VERBS AGREE IN NUMBER?

The proteolytic activity of extracts of X from these organs were expected to reach a high level should read

The proteolytic activity . . . was expected . . .

The subject of the sentence is the singular 'activity', not 'extracts' or 'organs', even though these nouns are closer to the verb than 'activity' is. Check that singular subjects have singular verbs and that plural subjects have plural verbs.

Compound subjects linked by 'and' should have plural verbs too, except when the two subjects are so closely linked that they form one idea ('bread and butter is good for you') (Gowers 1986).

When 'or' links alternative subjects of unlike number, either reword the sentence or make the verb agree with the subject nearest to it (see the last part of the next sentence, for example).

If the subject is a collective noun, such as 'council', 'team' or 'number',

Table 8.1 Checklist: grammar and technical style.

Verbs:
(a) Subjects and verbs agree in number?
(b) Auxiliary verbs correctly used?
(c) Participles attached to their subjects?
(d) Infinitives attached to their subjects?
(e) Passive voice appropriately used?
(f) Verb 'to be' used with restraint?

Nouns:
(a) Abstract nouns not over-used?
(b) Nouns not over-used as modifiers?

Pronouns refer clearly to a preceding noun?
Relative pronouns (that, which) correctly used?
Prepositions correctly used?
Definite and indefinite articles correctly used?
Comparatives clear?
Surplus negatives avoided?
Simplest words and simplest ways of writing used?
Words used with precision?
Sexism, racism, parochialism, and dehumanizing terms transformed or removed?
Punctuation: commas, colons, semicolons, hyphens, apostrophes, quotation marks, and exclamation marks correctly used?
Spelling correct and consistent?

Typography:
(a) Numbers, symbols and abbreviations correct and clear?
(b) Italicization/emboldening indicated?
(c) Capitalization correct?

Table 8.2 Grammatical terms: some definitions*.

Verbs and verb forms	Express action or a state of being
Transitive	Requires an object; may be either active or passive
Intransitive	Does not require an object ('The professor talked')
Auxiliary	Modifies another verb ('be' and 'have' are the most common auxiliary verbs in English)
Participle	A non-finite verb form, either present ('working', 'going') or past tense ('worked', 'gone'), used adjectivally and in forming some compound terms
Gerund	A noun form of a verb; always ends in -ing ('Working is a hobby to them')
Active voice	Indicates that the subject acts upon the object of the sentence ('Smith set up the experiment')
Passive voice	Indicates that the subject of the sentence is operated on by the object of the sentence ('The experiment was set up by Smith')
Nouns	Names
Common	Name of a thing ('car')
Proper	Name of a person or place ('Fred', 'Paris')
Abstract	Name of something that has no material existence ('excellence')
Concrete or material	Name of something that has a material existence ('house')
Adjective	Modifies a noun ('small car')
Adverb	Modifies or qualifies a verb or adjective; often ends in -ly ('The bell rang loudly')
Preposition	Shows the relationship between nouns in a sentence; often has something to do with space or time
Parts of a sentence	
Subject	The noun(s) or noun substitutes in a sentence about which the main verb says something
Predicate	The main verb of a sentence; must be a finite verb, not an infinitive or a participle
Object	The noun(s) or noun substitutes that are acted on by the main verb of a sentence
Clause	A group of words that includes a subject and a predicate (a single independent clause is also a sentence)

* Based on Roberts (1987).

use either a singular or a plural verb, depending on whether the subject as a whole or its individual parts or members are being emphasized:

The council meets once a year

is correct because the council can be considered as a single body, but so is

The council were not in agreement

because it takes two or more to agree, as Fowler (1965) points out. If the word 'number' is the subject, treat it as singular when it comes after the definite article ('The number is high') and (usually) as plural when it comes after an indefinite article ('a' or 'an'), as in 'A number of changes have been made'.

(b) ARE THE AUXILIARY VERBS CORRECT IN A SERIES OF PASSIVE VERBS?

The different parts of the auxiliary verbs 'to have' and 'to be' can be omitted in a sentence containing two or more passive verbs – but it is not always correct to omit them:

The valve was closed and several washers removed for examination

is wrong because 'washers', the subject of the second verb, needs the plural auxiliary verb 'were', not the implied singular 'was'.

The valves were closed and the largest washers removed for examination

is correct, strictly speaking, but if the second auxiliary verb is left out when both subjects in such a sentence are plural, the reader may get the wrong impression:

The valves were closed and the largest washers examined

leaves the reader wondering for a moment whether the washers did the examining (change the sentence to 'We closed the valves and removed the largest washers for examination').

(c) ARE PARTICIPLES ATTACHED TO THEIR SUBJECTS?

It is easy to confuse the subjects of participles, especially when the participles have third-person subjects:

Studies on system Y designed to counteract virus X were reported in January (Smith & Jones 1990) is ambiguous because the subject of the past participle 'designed' could be either the studies or system Y. To remove the ambiguity the sentence can be reworded as

They reported studies on system Y that were designed to counteract virus X

or as

They reported studies on system Y, which was designed to counteract virus X.

In this example

Studying [present participle] these areas, the conclusions were obvious

the phrase 'Studying these areas' is again grammatically unattached (or dangling or hanging). The phrase appears to belong to the subject of the main clause ('the conclusions'). The solution is to substitute an active verb for the participle and name the subject:

When we studied these areas, the conclusions were obvious.

Check all words ending in 'ing' in the text and see whether they need a subject. If they do, is that subject correctly and unambiguously identified?

Unfortunately, the verbal nouns called gerunds have the same 'ing' ending as the present participle in English. Unlike the present participle a gerund can be the subject of a sentence:

Obtaining [gerund] the new premises allowed us to rehouse the animals

is correct but

Before obtaining [participle] the new premises the animals were housed in constant-temperature chambers

is wrong because the subject is grammatically ambiguous – though in this example the reader should be able to deduce that the animals did not go out and buy the new premises themselves.

101

(d) ARE INFINITIVES ATTACHED TO THEIR SUBJECTS?

Dangling infinitives are just as misleading as dangling participles. The (unstated) subject of an infinitive may be confused with the subject of the clause that follows the clause containing the infinitive:

To examine this theory substance X was first weighed

is wrong because 'substance X' is not the subject of 'to examine'. Adding 'In order' at the beginning of the sentence doesn't improve matters because the infinitive is still unattached. A correct version of that example is:

To examine this theory we [or they] first weighed substance X.

(e) IS THE PASSIVE VOICE OVER-USED?

Many journals now encourage authors to use the first person, active voice, as often as possible. That is, they prefer

I [or We] tested the hypothesis by doing experiments on X, Y and Z

to

The hypothesis was tested by experiments on X, Y and Z.

Over-use of the passive voice quickly sends readers to sleep. The active voice is shorter, clearer, and often more correct. Write 'I [or We] think' rather than the passive 'It is thought'. If you use 'It is generally thought' make it clear who had this thought. Don't, however, go to the other extreme and use 'I' or 'we' in every second sentence.

(f) IS THE VERB 'TO BE' OVER-USED?

Over-use of the verb 'to be' in its various forms leads to verbal anaemia. '["To be"] carries no freight, moves no spirits, packs no punch, hits no nail, arouses no enthusiasm, all things that only the good old "transitive" or active verb can do.' (Steinberg 1985). If you find you have used the various forms of 'to be' too liberally in your draft, substitute active verbs as often as you can – but remember that the verb has many legitimate uses.

Use 'there is', 'there are', 'there was' and 'there were' as little as possible; they are often over-used or clumsily used:

There were six areas under study

would be better as

We studied six areas

or as

Six areas were studied.

Nouns

(a) ARE ABSTRACT NOUNS OVER-USED?

Over-use of abstract nouns, like over-use of the passive, tends to make readers fall asleep.

The conclusion that the reduction of x to y was brought about by the addition of a and b to the mixture was published by Smith & Jones in 1989

is weighed down with abstract nouns ending in '-ion'. The abstract nouns in this sentence are longer than their corresponding verbs, 'conclude', 'reduce', and 'add'.

Sentences full of abstract nouns also tend to be full of 'ofs' and 'thes'. If you liberate the active verb hidden in every abstract noun these redundant words will disappear:

Smith & Jones (1989) found that x was reduced to y when they added a and b to the mixture.

When you see a weak past participle such as

occurred, effected, brought about, achieved, produced, carried out, conducted, done, performed

look for the abstract noun that often accompanies such a participle and see whether you can substitute an active verb.

No exploration of this possibility has yet been carried out

for example, will be more easily understood if you change it to

Nobody has explored this possibility.

(b) ARE NOUNS OVER-USED AS MODIFIERS?

Nouns can legitimately modify other nouns but long strings of modifiers (nouns, or nouns and adjectives) are often difficult to understand. Non-specialists may find phrases such as:

a steroid-induced GABA channel burst duration prolongation

completely impenetrable. Insert verbs or prepositions between groups of three (or at most four) nouns, or nouns plus adjectives, as in:

a steroid-induced prolongation of the burst duration of GABA-activated channels.

In sentences with too many abstract nouns, 'of' and 'the' may be redundant (p. 103) but in word strings you may need to insert these short words to make your writing clearer and more precise. Remember, however, that two or three words can sometimes be regarded as a single name. The phrase 'glutamate receptor subtypes' is much less awkward than 'different subtypes of receptors for glutamate': too much unravelling of strings of modifiers may seem ludicrous.

Pronouns

Pronouns stand in place of nouns and must refer unambiguously to their antecedents – the nouns or clauses they replace and that appear before them. Ensure that every pronoun, particularly 'it' or 'this' at the beginning of a sentence, refers clearly to an antecedent. Repeat a phrase rather than risk being misunderstood. What does 'This' stand for in the second sentence below?

In older strata the effect is different and may be found throughout the layers, but not in X. This is not a boundary, however, between two types of layer.

The relative pronouns 'that' and 'which' have distinct meanings in defining and non-defining clauses (a defining clause limits what the antecedent refers to; a non-defining clause does not limit the antecedent). The defining clause ('that ... applied') in the next sentence limits the subject of the sentence to the particular effect that appeared under those conditions:

The X effect that appeared after these pressures were applied was clear.

In the next example the non-defining clause ('which . . . applied') gives extra but not essential information and does not restrict the subject of the sentence:

The X effect, which appeared after these pressures were applied, was clear.

In the first of those two sentences 'which' could be substituted for 'that' without changing the meaning. In the second sentence 'that' could not be substituted for 'which' (see also under 'Punctuation'). Strict grammarians say that 'that' should always be used in the first kind of sentence – but this is one of those rules you shouldn't exhaust yourself observing.

Prepositions

Prepositions are used to show relationships between nouns. Prepositions in English include 'at', 'in', 'for', 'on', 'by', 'of', 'from', 'to', 'than', 'through', and many others. Most verbs can be used with more than one preposition, depending on the intended meaning, but some verbs take only one preposition ('I am tired of this picture', not – in normal usage – 'I am tired with this picture'). If you are not sure which preposition to use, consult a large dictionary or a grammar textbook.

Don't twist sentences round ('This is an affront up with which we shall not put') to follow the old rule that prepositions should not appear at the end of sentences.

Don't use 'different than' if you can use 'different to' or 'different from'. 'Different than' is acceptable only in sentences such as

The specimen has a different appearance now than it had last week

where it allows you to avoid the pedantic 'different . . . from that which' or the incorrect and ugly 'different . . . to [or from] what it had'.

The word 'following' sounds like a dangling participle when it is used as a preposition – as in 'The eggs hatched following incubation'. 'Following' can usually be replaced by 'after', which is unambiguous. 'Following' is correct when it is used as an adjective, as in 'The following recommendations'.

Articles ('the', 'a', 'an')

Articles are of two kinds, definite and indefinite. Writers whose first language is not English often mix them up, or leave them out, or put them in when they are not needed. Which article to use depends on whether you are writing about something specific or something more general.

Use the definite article ('the') to show which particular item you mean:

I did three experiments: the first one shows . . . , the second shows . . . , and the third is still under way.

Use an indefinite article, 'a' or 'an', when you don't need to specify a particular person or thing:

A student must attend many lectures.

Whether 'a' or 'an' is needed depends on the accepted pronunciation of the first syllable of the following word or abbreviation: 'a messenger', 'an enemy', 'a eutropic lake' – but write, for example, 'an mRNA', because the reader is likely to pronounce this (mentally or out loud) as 'an em R N A', not 'a messenger R N A'.

Which sort of article to use, or whether to use an article, depends on the sort of noun that follows the article. Use an indefinite article when the noun is generalized:

He built a model on which he could base his experiments.

Use the definite article before a noun that names something that has already been mentioned:

He built the models we had discussed earlier

but don't use an article before a plural noun that has not yet been referred to:

He built models on his day off.

Don't use articles before a proper or abstract noun in sentences of this kind:

Dublin was a beautiful city

Movement of this kind is unusual.

But use an article if you use such nouns as common nouns:

The Dublin I knew as a student has gone

The movement was a sudden one.

Comparatives and negatives

If you write

This experiment is more difficult and we will therefore need more time

will readers know what you are comparing the experiment with? Always ask 'than what?' when you write 'more', 'less', and so on, and ask 'relatively to what?' when you write 'relatively' ('The sample weighed relatively little').

Don't use too many negative terms in the same sentence. Two negatives are acceptable when they refer to alternatives, as in 'neither X nor Y was found' or 'They did not see X, nor did they see Y'. But in grammar, as in arithmetic, two negatives usually add up to a positive ('not uncommon' means 'common'). Negative expressions may also make sentences difficult or impossible to understand:

Only in this department was there insufficient material to prevent the work being done

is guaranteed to puzzle readers.

Verbosity and pomposity

Make every word earn its place.

Long words and complicated sentences are not essential features of good scientific writing, although they are often thought to be so. The best writing in science, as elsewhere, is simple, clear, precise, and vigorous. Decide what you want to say and say it as simply, informatively, and directly as possible. Write

California suffered x earthquakes between 1969 and 1989

not

A localization that has suffered from many perturbations of the earth's crust in the last two decades is California.

Amongst its other failings that example is a 'backward-running' sentence – one with its subject at the end when it would have been better in the normal place at the beginning.

Don't be so afraid of committing yourself to a clear statement that you introduce layer upon layer of 'hedging'. The sentence

107

It may seem reasonable to suggest that these effects may possibly be attributable to the presence of substance X

contains five layers of hedging, yet it simply means

These effects may be caused by X.

Don't suddenly hedge or retreat after a positive phrase; the sentence

This finding strongly suggests that Y may play no part in Z

leaves readers wondering what the writer meant.

Don't use 'candidate' as an adjective ('these candidate effects may').

English sentences tend to keep subject and verb quite close together. They also tend to be shorter than sentences in many other languages, though you might not believe this after reading some of the prose in scientific publications. Shorten and simplify your sentences whenever you can do so without harming your meaning. Try to keep most sentences to between 10 and 25 words, with a maximum of 40 words. But vary the length – a series of 10-word sentences is as monotonous as a series of 60-word sentences.

Aim to have a single idea in a sentence: different ideas belong in different sentences.

Choose short or common words in preference to long or archaic words: use 'use', not 'utilize' or 'employ'; 'before', not 'prior to'; 'after', not 'following', and so on (see also Appendix 2).

Cut out clichés and meaningless introductory phrases such as

In this connection we may say that . . .

On the basis of the data presented by others and of the findings of our laboratory . . .

Under prevailing environmental conditions . . .

Change

Studies some years ago by Braun & Groen (1982) showed that X eats Y

to

Braun & Groen (1982) showed that X eats Y

or – better, because 'showed that', 'found that', 'reported that', etc. are usually redundant – to

X eats Y (Braun & Groen 1982).

Remove unnecessary adjectives and adverbs, especially vague qualifiers such as

very, quite, rather, fairly, relatively, comparatively, several, much.

'Very' reduces the impact of the term you are trying to strengthen. Don't use 'relatively' unless you are relating one number or quantity to another.

Don't repeat the same idea in different words ('arrived one after the other in succession', 'swam through the water').

Avoid terms such as 'of interest': readers can decide for themselves what is interesting.

Imprecision

If you use simple words instead of elaborate terms you will avoid the further pitfall of using words imprecisely. Aim to use words that are exactly right for the context, just as you aim for absolute accuracy in your experiments or observations. Use a thesaurus to find the right word, if necessary, but don't be tempted into 'elegant variation' by the near-synonyms in the thesaurus.

If you are not sure of the exact meaning of a word, check it in an unabridged dictionary. Check it also in a dictionary from English into your own language if your mother tongue is not English. If you translate the text into English yourself, try looking up an important word in a dictionary from your own language into English; then check the English word in a large English dictionary, and finally look up the English word you select in a dictionary from English into your own language. You may have to repeat the process several times before you find the right word. If you are uncertain about a word or phrase, put the original in square brackets after your translation.

British journals often use the *Shorter Oxford English Dictionary* as their authority for spelling, but editions of the *Shorter OED* published to date (1991) don't provide up-to-date definitions of technical terms. The *Concise OED* is more up to date, with new editions published every few years (eighth edition 1990), but it doesn't provide enough definitions of technical terms. Instead, consult *Webster's third new international dictionary* or the largest science dictionary you can find, or a recommended dictionary in your particular branch of science.

It is easy to use even simple words in the wrong way if you don't check their meaning in a dictionary. For example, 'affect' and 'effect' are not interchangeable. Neither are 'content' and 'concentration' – and 'level' is not a good substitute for either of them. 'Comprise' does not mean 'constitute'.

Some words and phrases are misused, even in reputable journals. Words ending in '-ology' usually refer to the body of knowledge about a subject, yet in medicine words such as 'pathology', 'morphology', and 'aetiology' are often used in sentences of this kind:

The pathology was found to be due to Y.

This should read

The abnormality [or lesion, or damage] was due to Y.

Some words become imprecise if they are used in the wrong place in a sentence. One of these is 'both', which is often redundant, so can be left out. Another is 'only', which can be used as either an adjective or an adverb. To use 'only' correctly, put it as close as possible to the word it qualifies:

He loved only her

He only loved her

Only he loved her.

Avoid euphemisms (inoffensive words substituted for words thought to be offensive). Use 'kill' not 'sacrifice', 'died' not 'passed on', 'the rats were starved for x days' not 'food was withheld for x days'.
Write 'In 1991 we . . . ', not 'This year we . . . '.

Sexism, racism, parochialism, and dehumanizing terms

Sexist language is imprecise as well as undesirable. It can lead to such absurdities as 'Man breastfeeds his young'. Use terms that apply to both sexes ('humans', not 'man'; 'humankind', not 'mankind'; 'staff', 'personnel', or 'work force', not 'manpower'). Avoid using 'he' to represent both sexes, but don't over-use 'he or she' in trying to get round this problem, and never write 'he/she' or 'him/her'. Instead, write sentences in the plural:

Scientists are busy people

110

not

The scientist is a busy man.

Avoid using 'Miss', 'Ms', or 'Mrs': the name alone is sufficient. Take care, however, to refer to female scientists as 'she' rather than 'he': it is the assumption that all scientists are men that gives offence. If you can't tell from the name whether someone is a man or a woman, use 'Dr' until you find out which sex the person is.

Make sure that you use terms that are both precise and acceptable for the names of races. 'Native American' is now preferred to 'American Indian'; 'Oriental' is not the name of a race, and it doesn't include Asian Indians. In medical papers do not mention the race or colour of patients ('Caucasian', 'black') if these have no obvious clinical implications.

Don't use disparaging terms such as 'backward nations', 'primitive societies', or 'imperialist warmongers'.

Don't use expressions such as 'in this country' (which country?), 'foreign' (foreign to whom?), or 'our Scottish salmon'.

Don't use terms such as 'subjects', 'cases', 'amnesiacs', 'geriatrics', and so on when you are referring to people; these words are dehumanizing as well as imprecise. Instead, use 'volunteers', 'patients', 'patients with amnesia', 'patients in geriatric wards', and so on, as appropriate.

Punctuation

Punctuation presents numerous problems but if you keep most sentences short and simple you will avoid most of the pitfalls. If your meaning is clear, a copy editor will be able to correct the punctuation for you, if necessary. Follow the advice in this section and then punctuate according to whatever rules of English usage you already know. The section on 'stops' in Fowler (1965) and those on punctuation in Huth (1987) and the *CBE style manual* (CBE Style Manual Committee 1983) offer general help, if you need it, and Carey's book *Mind the stop* (1971) provides more detailed discussion.

(a) COMMAS

Never separate the subject of a sentence from its verb with a single comma. Use either two commas for a parenthetical clause, or no commas:

The specimens, each of which was cruciform, weighed 90–100 g

and

The specimens were cruciform and weighed 90–100 g each

111

are both correct.

Make sure you have punctuated adjectival clauses unambiguously:

The specimens that weighed over 10 g were collected from X

means that specimens that did not weigh 10 g were present (but were not collected). The adjectival clause 'that weighed over 10 g' is called a defining clause because it separates the 10-g specimens from the others, whereas

The specimens, which weighed over 10 g, were collected from X

means that all the specimens weighed over 10 g. The non-defining or commenting clause in this version doesn't separate the specimens into categories. Use 'that', or sometimes 'which', without commas, for a defining clause. Always use 'which', with a pair of commas, for a commenting or non-defining clause (see 'Pronouns', earlier in this chapter).

Use a comma before the 'and' or 'or' at the end of a series of three or more items if you are writing for an American journal ('apples, oranges, and bananas'). Most but not all British journals omit this 'serial comma'. Even in British English, however, the comma is sometimes essential to avoid ambiguity: 'green apples, red oranges and bananas' would mean that the bananas were red too. (The serial comma is used in this book to comply with the publisher's 'house style'.)

If you use several adjectives before a noun, putting commas between them is, strictly speaking, a matter of choice (Gowers 1986, p. 165), though most publishers now prefer fewer commas rather than more. Don't put a comma before the final adjective of a series if it describes the species to which the noun belongs or is more closely related to the noun than the other adjectives are:

A large juicy red apple

or

A large, juicy, red apple

but

A large, juicy, red Cox's apple.

(b) SEMICOLONS AND COLONS

Use a semicolon to separate closely related clauses or to separate parts of a list when the parts already include commas; use a colon to introduce a list or before a clause that contrasts strongly with the preceding clause, or before a clause or phrase containing a climax or conclusion. A colon is stronger than a semicolon, which in turn is stronger than a comma.

(c) HYPHENS

Hyphens seem to be going out of fashion, although they do a useful job. Some journals forbid hyphens even in compound terms where they can be used to separate certain sequences of letters ('re-establish', 'freeze-dry'). Provided that your target journal has not banned them, use hyphens to clarify meaning, as in

An activity-mediated competition for this factor plays a role

or

And live alone in the bee-loud glade (W.B. Yeats).

Lack of a hyphen in sentences like the next two examples either confuses readers or amuses them:

They studied random samples by means of a doctor administered questionnaire.

Too late for fat lovers to repent (*New Scientist*).

A hyphen after 'doctor' in the first example and 'fat' in the second would have clarified the meaning.

Use a hyphen in words such as 'un-ionized' (not ionized), 're-form' (form again) or 're-sign' (sign again), to distinguish them from 'union-ized' (having a trade union), 'reform' (improve) or 'resign' (give up one's job).

Hyphens are rarely needed between adverbs and the words they qualify. Use them only if ambiguity is possible.

A beautifully designed experiment

needs no hyphen but

a little used car

is ambiguous, as Fowler (1965) points out.

(d) APOSTROPHES

Apostrophes seem to be going out of fashion too. They may be used to indicate possession ('nature's silyl group') or elision of a letter ('can't') but are best kept to a minimum in scientific writing. Don't confuse the possessive 'its' (no apostrophe) with the contraction 'it's' ('it is').

When names end in 's', add an apostrophe and a second 's' for the possessive ('James's results'). Don't use an apostrophe for the plurals of letters ('Ps and Qs'), dates ('the 1990s'), or words ('whys and wherefores').

(e) QUOTATION MARKS

Follow journal style for quotation marks, and see Ch. 9, 'Quotations and quotation marks'.

(f) EXCLAMATION MARKS

In scientific papers use exclamation marks as mathematical symbols only.

TECHNICAL STYLE

As well as revising the literary style of your draft you must check the technical style, that is spelling and the use of capital letters and italics (usually indicated by underlining). Inconsistency in these matters may not worry the editor or referees as much as other kinds of inaccuracy – but neither will it prejudice them in your favour. Pay special attention to tables and figures, which are expensive to put right at the proof stage. (See also Ch. 9.)

Spelling

Have you spelt the same word in the same (correct) way throughout? Some words can legitimately be spelt in two ways, neither of which is necessarily better than the other ('neuron', 'neurone'), but you should use one form consistently – preferably the version usually used in your target journal. Compile a word list: every time you take a decision on spelling, record it, and check your second draft for conformity to the list.

The choice in spelling is sometimes between British and American versions of a word. Consult a recommended dictionary to see whether one form is preferred to another, or to check which version is British and which is American. If possible, use a dictionary such as *The Random House College dictionary* or *Chambers 20th century dictionary* that gives both British and American spellings. Many journals accept and use both British and American spelling, provided that only one of these is used throughout an

Table 8.3 Some differences between British and American spelling*.

British	Example	American	Example
-re	metre, fibre	-er	meter, fiber
-ce	licence, practice (nouns only; verbs spelt with -se)	-se	license, practise (nouns and verbs)
-our	tumour, honour	-or	tumor, honor
-ogue	catalogue	-og	catalog (but vogue)
-amme	programme (but computer program)	-am	program
-ey	grey	-ay	gray
-l	fulfil, instal	-ll	fulfill, install
-iset, -yse	hospitalise, analyse	-ize, -yze	hospitalize, analyze
em-, en-	embed, ensure	im-, in-	imbed, insure
-ae-, -oe-	anaemia, oedema	-e-	anemia, edema
-oeu-	manoeuvre	-eu-	maneuver
-ll-	traveller	-l-	traveler
-eing	ageing	-ing	aging
-ph-	sulphur	-f-	sulfur (but telephone)

* Note that this list is not complete and that the examples do not represent invariable rules. The only safe rule to follow is to look up every word you are not sure about in an appropriate dictionary.
† The Shorter Oxford English Dictionary prefers -ize to -ise. The version used in British books and journals depends on the publisher's house style (-ize is used in this book, for example).

article. Others use only one kind of spelling throughout the journal. See Table 8.3 for some of the main differences between the two kinds of spelling. Use a spelling-checker if you have one on your word processor, but find out whether it uses British or American spelling. And remember that the program won't point out words that are wrong but correctly spelt ('from' when you meant 'form', or 'casual' instead of 'causal').

Typography

Follow the journal's instructions and usage in your discipline for most typographical details. Try to write numbers, symbols, and abbreviations in the form and spacing approved by the journal ('1 mg' or '1mg'?).

Underline any words or symbols in your paper that should be in italics in the printed version, or put a wavy line below any that should be in bold type, or use the appropriate codes in a computerscript. Do not underline words for emphasis; instead, write the sentence in such a way that readers will realize what you are emphasizing. Check whether words that need initial capitals are clearly indicated if you are not going to retype the draft yourself. Some advice on italicization and capitalization is given below, but first check the journal's instructions on points like these.

In mathematical work, mark scalar variables for italic type and vectors for bold type. Avoid using superscripts or subscripts that have to be placed directly above or below variables. Avoid using multiple bars or tildes above or below variables. Keep the use of second-order superscripts or subscripts to a minimum; the second level runs into the line above or below and makes printing difficult or inelegant, or both.

Use initial capitals for points of the compass when they are part of a formal name, such as 'North America', or refer to a recognized region ('in the West', 'unemployment in the North'), but not if they are used in a general way ('in southern England'). Use initial capitals for other formal names ('the River Thames', but 'the Caspian sea'). In geology use initial capitals for formal stratigraphic names ('Cambrian System', 'Upper Chalk') and for the names of structural features qualified by a place name ('Glarus Nappe'). Don't use initial capitals for informal stratigraphic names ('chalky formation') unless these are used as part of a formal name. Names of soil types should be in lower case ('brown earth'). Soil horizons are denoted by capitals, with qualifying suffixes in lower case, closed up and on the line ('Bg', 'Cfe'). Lithological discontinuities are denoted by numerical prefixes ('2Bt', '3Cg'). In radiocarbon dating use 'bp', 'bc', 'ad' for uncalibrated dates and 'BP', 'BC', 'AD' for calibrated dates.

In the life sciences underline genus names and give them an initial capital only. Use underlined lower-case for species names. Don't underline 'sp.' and 'spp.' ('Glechoma hederacea'; 'Polysiphonia spp.'). Write common names of plants, animals, and so on in lower case, unless proper nouns form part of the name ('ground ivy', 'blue whale', 'Oxford ragwort'). Use initial capitals for names above genus names, provided that these are used formally ('Cyprinidae', but 'a cyprinid'). Write Latin names for parts of the body in ordinary type (roman, not underlined).

Underline conformational prefixes ('cis', 'trans', 'p-', 'c-', 'm-').

In chemistry mark or code 'O' as a capital letter whenever it is used for oxygen and mark 'l' as the lower-case letter 'el' whenever it is used in 'Cl' and 'Al', for example. Don't put spaces before or after chemical bonds. Use an initial capital letter for trade names, not quotation marks.

OBTAINING COMMENTS ON THE SECOND DRAFT

When you have revised the paper structurally and stylistically, retype it and give copies to any co-authors. If possible, show it also to colleagues in the same or related fields of work and to a friend in a different discipline. This kind of review will help to prevent later delays in your target journal's reviewing process, because readers unfamiliar with the manuscript are more likely than you are to spot inconsistencies, jargon, lapses in logic, and other faults. Give everyone complete copies of the paper, including the abstract, tables, figures, legends, and references, because these are essential parts of the paper. Ask for comments in writing, for preference; this usually saves time and avoids misunder-standings.

SUMMARY

(1) Make sure that you have used the correct person, voice, and tense, and that you have removed all unnecessary jargon;   (2) Check that the second draft fulfils the grammatical and technical requirements listed in Table 8.1;   (3) Retype the draft and obtain comments from any co-authors and from suitable colleagues.

CHAPTER NINE

Preparing the final version for submission

Typing: general advice   Materials and format   Pagination   Title
page   Text   References   Checking and correcting the typing
Obtaining final criticisms

When you have revised the paper structurally and stylistically make sure
that the final versions of the tables and figures (Ch. 3 and 4) are ready.
Then type or retype the paper, or correct it on your word processor/
microcomputer, in the ways described in this chapter – but with the
journal's instructions taking precedence, as always. Some of the advice
here may be useful for earlier drafts as well as for the final version.

TYPING: GENERAL ADVICE

When you prepare the final version bear in mind that editors and referees
(or 'reviewers') get their first impression of a paper from its physical
appearance. Although scientific merit and suitability for the target
journal are (one hopes) the main criteria for acceptance, a well-presented
article will be easier for the editor and referees to deal with than a badly
typed bundle of scrap paper. Making the job of reading the manuscript
easier for the referees, who are busy people rarely paid for assessing
other people's work, is also a matter of elementary courtesy. A good
appearance may influence these judges in your favour. If your article is
typeset rather than printed directly from your manuscript or disk the
same qualities will help the copy editor and the typesetter.

   If someone else is doing the typing or word-processor keyboarding for
you, make sure that the manuscript is as legible as possible and that the
format (see next section) matches the journal's requirements. If necessary, make a list of instructions on how to type the paper for your target
journal. If you want the manuscript or printout ready by a certain date,
tell the typist in good time and say whether you are asking for a draft or
for what you hope will be the final version.

If the journal is produced directly from authors' manuscripts (camera-ready copy), your manuscript must be prepared to a high standard. The journal will provide detailed typing instructions and may supply special paper with grid lines printed on it to guide the typist. The journal will also give special instructions if the optical character recognition (OCR) method is used, in which authors' manuscripts are electronically scanned to obtain a machine-readable record that is later used for typesetting.

If the journal is typeset directly from authors' floppy disks or magnetic tapes (that is, from computerscripts) you may be asked to code the computerscript before submitting it. More probably, though, an edited printout will be returned to you after acceptance and you may be asked to transfer the editorial additions and changes to your computerscript at that stage. If you are asked to add typesetting codes too, the journal will send you special instructions and you may be asked to supply a fresh printout as well as a corrected and coded computerscript. Coding straightforward text is easy to do but if your material is complex you may need to talk to the editorial staff about who should do the coding.

If the journal is produced in an electronic form only (rare at present), the instructions to authors will tell you how to format the material and how and when to transfer it to the electronic system.

MATERIALS AND FORMAT FOR THE MANUSCRIPT OR PRINTOUT

(1)   Use white paper of at least $80 \, g/m^2$ and A4 size (about $210 \times 297 \, mm$) or, in North America, 50 pounds and 8.5×11 inches (about $215 \times 280 \, mm$).

(2)   If you are using continuous (fanfold) paper, separate the pages when the manuscript has been printed and remove the sprocket-hole strips from the edges of the paper. Make sure that fanfold paper is of the same quality and size as recommended in (1) above.

(3)   Use an ordinary typeface (e.g. Courier or Prestige on a typewriter, Times Roman/Dutch on a computer printer) with 10 (pica) or 12 (elite) characters to the inch or 2.5 cm. Don't use italic or other non-standard typefaces.

(4)   If you are using a word processor/microcomputer, use a printer with high quality output, including a typeface that has distinct ascenders and descenders (the strokes above and below the line of text, as in the letters b and p). Some line-printers and dot-matrix printers provide output that no one other than the author can bear to read.

119

(5)   In manuscripts, but not in computerscripts, indent the first line of each paragraph by, say, five spaces. Alternatively, mark the beginning of each paragraph by hand with a paragraph mark (see Fig. 11.1). If all paragraphs begin at the left-hand margin (flush left) the copy editor or typesetter cannot always tell whether a new sentence at the top of a page starts a new paragraph or runs on from the previous paragraph.

(6)   Change the typewriter or printer ribbon as soon as the type begins to look faint.

(7)   Print or type EVERYTHING (text, tables, legends, reference list, footnotes) in double spacing – that is, with a blank line after every typed line – to give about 30 lines to the page. Do NOT try to save paper or meet page limits by using small print or leaving only half a line empty after each typed line; 1½-line spacing, widely used in continental Europe, doesn't give British or American editors or copy editors the space they need to make linguistic or other changes or to mark instructions for the typesetter.

(8)   Leave an extra line above and below headings, equations, and formulae, except when the instructions for typing a computerscript ask you not to do this.

(9)   Leave margins of at least 25 mm all round, with 35 mm or more on the left, to give the editor and copy editor room for writing queries to the author or instructions to the typesetter. In a computerscript start typing at column 1 (set the left-hand margin appropriately at the printing stage instead of on the screen).

(10)   Leave the right margin unjustified (uneven, or 'ragged right') unless the journal specifically asks for the lines to be justified. Justified lines make it harder for the editor/copy editor to count the words and to check for typing errors.

(11)   Don't hyphenate words at the ends of lines; if your word processor hyphenates words automatically, switch it to its unhyphenated word-wrapping mode.

(12)   Arrange the pages in the way described below ('Pagination) and (if they haven't already been numbered) number them, preferably at the top right-hand corner. Start with the title page and include all the parts listed below. Don't type page numbers or running heads in computerscripts that are to be used for typesetting, but make sure that numbers and running heads appear on the printout.

PAGINATION

Arrange the parts of the manuscript in the order shown below (items a to k) and start a new page for each part, unless the journal gives other instructions.

(a) Title page, laid out as discussed below.
(b) Abstract, with key words or indexing terms if the journal requires them to be placed with the abstract.
(c) Key words or indexing terms (if not required on the title page or with the abstract).
(d) Abbreviations or glossary, and a list of special symbols used, if any of these have to be listed separately.
(e) Text.
(f) Appendix, if you are including one.
(g) Acknowledgements.
(h) Reference list.
(i) Footnotes – if the journal asks for these to be placed on a separate page or pages – and endnotes, if any.
(j) Tables (each table on a separate page or pages).
(k) Legends for the figures, typed as a consecutive series (i.e. don't start a new page for each legend unless the journal requests this).

Inserting or removing pages

If you have to add an extra page to a completed typescript after page 5, for example, number the new pages '5A', '5B', and so on. Then write '5A follows' at the bottom of page 5, and write 'p. 6 follows' at the bottom of the last inserted page. If you remove a page, add its number to the number on the page before or after: if page 6 is removed, for example, make page 5 into '5 and 6'.

If a correction or insertion is too long to be written or typed neatly above the relevant line, or typed on a slip of paper which can be pasted over the original, retype or reprint the whole page and renumber any extra pages that are needed. Never type insertions on the back of a page or up the sides at a 90-degree angle to the rest of the text or on slips of paper that are then stapled or pinned to the page.

TITLE PAGE

Centre or left-justify the title of the paper in the top third of the page. Type the by-line – the names and institutional addresses of the authors – below the title in the form used in the journal.

List your co-authors in the order you have all agreed on (see Ch. 2, 'Agreeing the order of authors' names) and ask them exactly how they like their names written. Some people always write one given name in full (and some journals specifically request this). Other people prefer to use initials instead of given names if they have the choice, or they may leave out one or more initials. You should decide early in your career how you want your name to appear in by-lines and then write it consistently in the same way. If you vary the style (J. Smith, John Smith, John L. Smith) you will be listed in different places in sources of bibliographic information such as the *Science Citation Index* and the articles will appear to be by different people. The same problem arises for women who change their names on marriage, divorce, or remarriage. Consider whether to keep to your unmarried name for publication purposes.

If your co-authors come from several departments or institutions, use symbols or superscript characters to make it clear who works where, if this is not obvious from the by-line layout. If an author's address has changed since the work recorded in the paper was done, put the current address in a footnote linked to the author's name by a symbol or superscript character. Write the names and addresses of departments or institutions in the language used on their official stationery (transliterated if necessary).

Other information needed on the title page may include authors' degrees; job titles; key words; the name and postal address of the author to whom correspondence, proofs, and requests for reprints should be sent (if this is not clear from the by-line); sources of support for the work (see Ch. 5, 'Acknowledgements'); and the short title (see p. 69).

TEXT

Throughout the paper, including the title page, tables, and legends, use capital letters for words or abbreviations that are normally written in capitals (FORTRAN or DNA, for example), for the initial letter of the first word of a sentence, and for the initial letters of proper names, trade names, and so on. Don't use capitals for any other words. Underline (once) any words that are to be printed in italics, such as the Latin names of species. Don't underline the title of the paper, except for words or characters that are to be italicized. Don't use double underlining unless the journal specifically requests it (to indicate that small capital letters are needed). Too much underlining makes it difficult for the copy editor to mark manuscripts clearly for typesetting. If you want a word to be underlined rather than italicized in the printed version, write a note in the margin – 'underline, please' and circle this request. (See also Ch. 8, 'Typography'.)

Don't break long words by hyphenating them at the ends of lines, especially not words that already contain hyphens (see 'Materials and format', point 11, above).

When you refer to other parts of the paper, write 'see Methods section', for example, rather than 'see p. 5', because the page number will change later – and might be forgotten or changed to the wrong number during the production stage.

Headings

The manuscript should preferably have no more than four orders of heading, including the main headings (see Ch. 5, 'Headings'). Identify the order of headings by writing 1, 2, 3, 4 or A, B, C, D (encircled) in the margin. Type the headings in the style required by the journal; code them appropriately if asked to do so for a paper that will be typeset from a computerscript. Capitalize only those words or letters that are normally written in capitals, except of course when whole headings have to be typed in capitals. Don't leave spaces between each letter in headings or place full stops (periods) at the end of them. Don't underline headings unless you are sure they will be printed in italics, or unless the journal asks you to do so. Instead, set them off from the text by leaving an extra line before and after them (except in a computerscript, when the journal may ask for coding rather than spacing). Don't centre or indent headings in the final manuscript.

Ambiguous or unusual characters

If there could be any doubt about whether 'l' is a lower-case letter ('el') or a figure one, whether 'x' is a letter or a multiplication sign, and whether 'O' is a capital letter or a zero, write a note in the margin identifying these characters. If you are using a word processor, be sure to use the number keys for '1' and '0'. Don't write the name of a Greek letter such as alpha or beta in full if the convention in a particular discipline is to print 'α', 'β', etc. If your typewriter or printer doesn't produce Greek letters or other unusual characters, write them in by hand. Identify them by writing their names in the margin (encircled) the first time each appears in the manuscript.

Numbers and mathematical formulae

Type numbers, not words, for all quantities attached to abbreviations for units of measure (5 g, 10 m), except when journal style requires numbers to be written in full at the beginning of a sentence. If you or the publisher objects to numbers at the beginning of a sentence, rewrite the sentence so that the number falls elsewhere. When quantities mentioned in the text refer to something other than units of measure, spell out whole numbers from one to nine or whatever limit the journal may stipulate ('nine records'), and use numerals for numbers larger than that limit. But don't mix numerals and spelt-out numbers when the number refers to the same noun in a phrase or sentence (type '4 out of 15 regions', not 'four out of 15 regions').

Use a full stop for decimal points, not a comma (21.9, not 21,9), unless you are submitting an article to a journal that uses commas for the decimal point. Place a zero before the decimal point for all quantities between 1.0 and −1.0 (0.695; −0.28) in case the decimal point is overlooked. Do not use a raised decimal point unless asked to do so.

When numbers consist of more than four digits, leave a space between each group of three on either side of the decimal point:

5 213 504, for example, or 23.587 62

If numbers have no more than four digits to the left or right of the decimal point, type the digits without spaces (5213; 0.5876) except when you are aligning them with numbers of more than four digits in the columns of a table:

9 876
69 578

Always align columns of numbers on the decimal point. If ± or = signs are used in columns, align first on these signs and then on the decimal point, as shown on p. 29.

Type mathematical work carefully, as far as the symbols available on your typewriter or printer allow you to type it at all. If necessary, write mathematical expressions in by hand and name them in the margin. Special codes will have to be used for these if typesetting is done direct from a computerscript.

Don't use a solidus (slash) twice in the same expression, because this can be ambiguous. Instead of 5 W/m/K, for example, write 5 W/(m K) or $5 \text{ W m}^{-1} \text{ K}^{-1}$ or 5 W/m per K (Royal Society 1975).

Always use the decimal system in preference to fractions. Whenever possible write or transform mathematical expressions so that they can be printed on a single line.

Avoid using the symbol ‰ for 'per thousand' – it can easily be mistaken for %. Instead, give the unit of measure (5 ml l^{-1}) or write out 'per thousand' if there are no units. Use 15 mg l^{-1} or $15 \mu\text{g g}^{-1}$, not mg%.

In a few countries (Norway, for example) the ÷ sign is used as a minus sign, not a division sign. If you use this sign as a minus sign, make it clear what it means.

Abbreviations

Keep abbreviations in the text to a minimum (see Ch. 7, 'Nomenclature', abbreviations, and footnotes'). Any abbreviations you use must be either internationally acceptable or defined at first mention. As you type the paper make sure that these conditions have been met and that the abbreviations are used consistently. Once an abbreviation has been introduced and defined, there is no need to use the unabbreviated term again in a journal article of average length, except in figure legends and tables, which must be comprehensible without reference to the text.

Use recognized abbreviations, including those for SI units, for units of measure (A for ampere, C for coulomb, etc.) when a number precedes such units in the text. Note that the abbreviation for 'second(s)' is 's', not 'sec'.

Quotations and quotation marks

Copy passages or phrases from other people's work exactly as they appear in the original, including mistakes, if any (see Ch. 7, 'Accuracy and currency of citations and quotations'). Remember that permission is needed for quotations of over 100 words or 5% of the original article (see Ch. 2, 'Coping with copyright'). Indent long passages from the left margin, or code them appropriately in a computerscript. Use quotation marks for short quoted passages or phrases included in the running text (normal unindented text). For journals published in the UK use single quotation marks (if available on your typewriter or printer), with double quotation marks for a quotation within a quotation – unless the journal commonly uses double quotation marks. For American journals use double quotation marks first, with single marks for a quotation within a quotation. Don't use angle brackets (<< ... >>) for quotations.

Put the appropriate (single or double) quotation marks round a newly coined or unusual word, or round a word the first time it is used in an unusual way. It is not necessary to use quotation marks the second or any other time such words are used.

Hyphens, dashes, and minus signs

Differentiate between hyphens, short dashes ('en dashes' in typography), long dashes ('em dashes'), and minus signs if the copy editor is likely to have a problem identifying which of these is which. If no special style is laid down in the instructions to authors the guidelines that follow may help.

Don't leave a space before or after a single dash when it is used for:

(1) A simple hyphen – e.g. in a compound word such as 'by-line' or 'freeze-dry';

(2) An en dash – e.g. to indicate a range (20–25 mg), distance or movement (London–New York), or a combination (gas–liquid) (write 'en' or '1/N' lightly in pencil above these in a manuscript if you think the copy editor or typesetter might miss them);

(3) A chemical bond – e.g. C−H, C=O (write 'bond' in pencil above these).

Type two dashes without spaces before or after the dashes when they represent:

(4) An em dash (e.g. to separate or emphasize a portion of a sentence).

Put a space before and after:

(5) A minus sign or an equals sign (e.g. $10 - 7$, $10 - 3 = 7$).

Put a space before but not after:

(6) A negative number (e.g. -7).

If typesetting is done direct from a computerscript the journal may ask for special codes to be used for the different kinds of dashes as well as for chemical bonds, mathematical operators, and any other unusual characters.

Typing tables and legends for figures

See Chapters 3 and 4.

Typing and checking references

Type and check references with special care (see Ch. 6).

CHECKING AND CORRECTING THE MANUSCRIPT OR PRINTOUT

Read and check the manuscript or printout carefully: checking a printout is more efficient and easier on the eyes than checking text on a screen. If possible, get someone to read the previous draft aloud while you read and correct the new version. Use a brightly coloured ink for your changes and corrections, so that the typist or typesetter can see them easily.

Make changes and corrections in or between the typed lines, not in the margins, although on word processor printouts you should also make a mark in the margin beside each line containing a correction. Don't use proofreaders' marks and procedures when you correct a manuscript or printout. A typist can cope with corrections and instructions made in or above the line of type more easily than with corrections written in the margin. If you must place added or changed material in the margin, circle the material and draw a line from the circle to an insert mark in the text where the material is to go. If there are long inserts, follow the advice on p. 121 on inserting pages.

If you are asked to insert typesetting codes, check these very carefully indeed: one mistake and the rest of the paper may be typeset in italics or Greek or in weird symbols you may never have seen before.

Check all numbers in the paper very carefully too, and check that spelling, symbols, and abbreviations are consistent and correct throughout. (See Ch. 8, 'Spelling'.)

Check the headings and subheadings again for usefulness and consistency. Make sure you have referred to each figure and table at least once in the text and that each first mention is in sequence (not Fig. 2 before Fig. 1). Indicate the approximate position of figures and tables by a circled note in the margin beside the first mention of each one ('Fig. 3 near here', or 'Table 1 here').

If one or a few lines have to be corrected on a typescript that is destined to be camera-ready copy or OCR copy, use correction tape, or cut out the incorrect lines and paste in retyped sections – aligning them carefully on the page – or retype/reprint complete corrected pages. If you use correcting fluid, start a new bottle or use recently thinned fluid, use as little of it as possible, and let it dry completely before you type the correction. Alterations to camera-ready copy can look dreadful in print: retyping the page is preferable.

FINAL ROUND OF CRITICISM

When you have finished checking and correcting the paper, put it together in the order listed in the section on 'Pagination' above, including the tables and figures. Make enough copies of the complete corrected version for each of your co-authors, for the head of your department and for any person or committee in your institution from whom or from which you need final clearance before the paper is submitted or before it is published.

If you receive critical comments, decide on the final changes and make corrections in the ways just described. Make sure that your co-authors agree to the changes.

If (or when) you receive congratulations rather than criticisms, make the required number of copies of the manuscript or printout for the journal (see the instructions to authors), including a complete copy for yourself. Don't submit carbon copies, blurred photocopies, or copies on unbleached or tinted paper. Don't fold the copies. And don't exceed a page limit without good reasons for doing so.

Make sure that the manuscript you submit really is the final version, apart from later changes requested or suggested by the editor or referees. Don't submit a later 'improved version' containing your afterthoughts. This is totally unacceptable to editors and referees and wastes everybody's time.

Finally, go through the checklist here (Checklist 9.1) before writing to the editor of your target journal (Ch. 10).

Checklist 9.1   Items to check before submitting a paper[*].

---

(1) Text pages numbered consecutively, from the title page onwards, in each copy of the paper?

(2) Appropriate sets of tables, figures, and legends included with each copy of the paper?

(3) Title, abstract, and index terms appropriate?

(4) Address for correspondence included on title page?

(5) Tables and figures numbered consecutively in two series, in the order of their citation in the text, each identified with the author's name, and understandable without reference to the text?

(6) Table titles and figure legends refer accurately to the correct tables and figures?

(7) Marginal notes show where each table and figure is first mentioned?

(8) Each citation in the text, tables, and legends listed in the reference list?

(9) Each reference in the reference list cited at least once in the text, tables, or legends?

(10) Each citation in the reference list cited accurately, as shown by comparison with the original source, and in the form required by the journal?

(11) Each symbol for a footnote in the text has a corresponding footnote, and each footnote a corresponding reference symbol?

(12) All special requirements of the target journal met?

(13) Final version of paper read at least twice, once against the pages from which it was typed and once for a general view?

(14) Copies of releases, copyright form, letter of approval from ethics committee, and covering letter to the editor (Ch. 10) enclosed, as necessary?

---

[*] Adapted from the *CBE style manual* (CBE Style Manual Committee 1983, p. 31–32).

SUMMARY

(1) Prepare to type the paper to a high standard and according to the journal's requirements;   (2) Type the paper double-spaced, with unjustified right-hand line ends and without hyphenation; leave wide margins; use a standard typeface on good-quality paper of the recommended size and weight; use good-quality printer output, and separate fanfold paper;   (3) Arrange the parts of the manuscript in the recommended order, starting a new page for each part listed here, and number the pages if they haven't already been numbered; (4) Put the required information on the title page, listing your co-authors in the agreed order and writing their names in the way they prefer;   (5) Deal with matters of technical style (affecting spacing, capitalization, italicization, hyphenation, headings, ambiguous or unusual characters, numbers and mathematical formulae, abbreviations, and punctuation) in the recommended ways;   (6) Type tables and legends for figures as recommended in Chapters 3 and 4;   (7) Type references with particular care (see Ch. 6);   (8) Check the typescript or printout and make corrections in or between the typed lines;   (9) Obtain final criticisms from co-authors and others and make the final changes to the manuscript, first obtaining the agreement of your co-authors to these changes; (10) Make the required number of copies for submission to the journal.

CHAPTER TEN

Submitting the paper

Writing a covering letter   Mailing the manuscript   Checking the
manuscript's progress   Responding to the editor   Paper accepted
Revision requested   Paper rejected   Referees: general advice

When your manuscript is ready for submission write a letter to the editor
to go with the paper. Later, reply to letters from the editor as quickly as
possible, not only as a professional courtesy but also to speed the
processing of your paper.

WRITING A COVERING LETTER

Keep the covering letter short, simple, and to the point. Don't list all your
achievements to date or claim that you and the editor have mutual
friends or are distantly related. Don't ask a well-known scientist to
forward the paper to an editor on your behalf unless that scientist is
pre-reviewing the paper for a journal that uses a preliminary reviewing
system.

If the work you are reporting is part of a series or is closely linked to an
earlier paper in the same journal or elsewhere, mention the earlier
publications (see Ch. 1, 'Deciding what kind of paper to submit for
publication'). Some editors like to receive copies of the earlier papers. If
the work or part of it has been reported before in any form – for example,
in a conference abstract or in a newspaper or magazine – tell the editor
about the reported version and enclose a copy of it.

If you have a lot of tables or figures, or if they are very large, tell the
editor whether you are willing to have some or all of them placed in a
suitable archive instead of being printed with the rest of the paper. The
journal's instructions will tell you whether such an archive is available.

If the journal publishes several categories of contributions, say which
category your paper belongs to. If the instructions to authors ask you to
suggest the names of possible referees (reviewers), include the names in
your letter and mention anyone to whom the editor could turn for more
information.

... write a covering letter to the editor ...

If you are going to be away for some time, tell the editor when you are due back. Arrange for someone to open your mail and deal with any letters about publication. If correspondence is to be sent to someone other than yourself, give that person's name, address, and telephone and fax numbers.

If the paper describes investigations in humans or other animals with a central nervous system, confirm that the work was authorized by the ethics committee of your institution or another appropriate authority and enclose a copy of the letter of authorization. Where relevant, assure the editor that informed consent was obtained in accordance with ethical guidelines or that experimental animals were well treated and cared for (see Ch. 1, 'Three sets of preliminary questions', and Ch. 5, 'Methods').

Enclose copies of letters ('releases') from copyrightholders giving you permission to cite unpublished work or reproduce tables, figures, or text from previously published material (see Ch. 2, 'Coping with copyright'). If your target journal prints a copyright assignment form in each issue and asks authors to use this when submitting papers, enclose a signed copy of the form. If necessary, get your co-authors to sign the same form or copies of it. Note that some journals regard the submission of a manuscript as constituting an assignment of copyright if the paper is accepted for publication.

If you have had the paper translated into English by a non-scientist and have not been able to have the translation checked by a colleague whose native language is English, explain this to the editor.

A typical covering letter reads something like the one in Figure 10.1. See also Appendix 1, 'Submission of manuscripts'.

1 January 1999

Dr P Smith
Journal of Porcine Investigations
Blandings Hall
University of Loamshire
Loamtown
LO1 3UP

Dear Dr Smith

I enclose two copies of an article entitled

   Prevalence of hydatid cysts in pigs in Lower Slaughter

by A. James and J. Stone. We should be grateful if you would consider this
original paper for publication in the journal.

The work reported in this article extends the work described in our earlier
article, "Incidence of infection in pigs in Upper Slaughter" (J Porcine Invest
1998;10:99-101).

I also enclose a copy of a letter from XYZ Publishers giving us permission to
use Fig. 1 from a paper by Dr B. Green.

We look forward to hearing whether you can accept this article
for publication.

Yours sincerely

A. James (Dr)

Senior Lecturer

Figure 10.1   Sample covering letter for inclusion with a manuscript submitted to
a journal.

MAILING THE MANUSCRIPT

Check that the copies to be submitted are complete, if you haven't
already done so (see Checklist 9.1). Include the original (top, or ribbon)
copy of the manuscript or printout among the copies you submit. Don't
staple the pages of this original copy. If some sets of figures are
photocopies, mark them 'Not for reproduction'.

Put the figures between thin card or other protective covering. Attach
the covering letter to the top copy of the manuscript. Put everything in a

strong envelope or other secure wrapping that is just large enough to hold the contents – if the envelope is too large the contents can move around and eventually push their way out before the package is delivered. Seal the package carefully, address it clearly, and make sure it is marked for the appropriate postal class (first class, airmail, foreign airmail, etc.). Don't use 'printed papers' or equivalent cheaper postal classes that are likely to receive lower priority than the more expensive classes. When necessary, add a customs declaration form or sticker ('scientific manuscript, no commercial value').

Keep a complete spare copy of everything, including high quality photographs, in case the package is lost or damaged in the post.

If the journal makes use of authors' disks or magnetic tapes, send the disk or tape if it is required at this stage. Pack it securely and enclose the appropriate number of copies of the printout, plus the figures and your covering letter. The disk or tape, however, is more likely to be needed after the paper has been refereed and accepted for publication.

FOLLOWING UP THE PROGRESS OF THE MANUSCRIPT

Most journals acknowledge receipt of manuscripts promptly. The instructions to authors or the acknowledgement you receive may say how long the journal will take to tell you the fate of the paper. If you've heard nothing by a week or two after the promised time, write to ask whether a decision has been made. If the time it takes to give a decision has not been indicated, write to enquire about your paper six to eight weeks after the journal acknowledged receiving it. Don't phone to ask what is happening to it. Don't submit the paper to any other journal until you get a letter of rejection from your target journal (see Ch. 1, 'Deciding what kind of paper to submit for publication')

RESPONDING TO THE EDITOR

The eventual letter you receive from the editor will carry one of three messages: the paper has been accepted, or it will be accepted or reconsidered if you revise it as suggested, or it has been rejected.

(1) Paper accepted

Papers are seldom accepted outright. If your paper is one of these rareties, acknowledge the editor's letter of acceptance briefly, and sign and return the copyright form if one is sent to you at this stage (your co-authors may also be asked to sign this form). If a computerscript is returned to you for minor changes and corrections to be made or for coding to be entered, or if you are now asked to supply a coded computerscript, enter the corrections or coding (or both) according to the journal's instructions and check the newly typed material carefully. Return the computerscript and printout to the editor by the requested date. If a date is given for the arrival of proofs (Ch. 11) check your diary to see whether you will be there to receive them. If you are due to be away for more than a few days and don't want the proofs to follow you to a conference or on holiday, ask someone else to check them for you. Give the editor this person's name and address, even if the information is also given on the title page of the paper.

(2) Revision requested

The letter you are most likely to receive is one asking for specific changes to be made before the article can be either accepted or reconsidered.

If the editor says that the paper will be accepted if the changes are made, consider the suggested changes carefully. If you agree they will improve the paper, make the necessary alterations. Make sure that your co-authors approve of the changes. Retype any heavily corrected pages before you return the paper to the editor. Return the corrected pages as well as the retyped copies. In the covering letter you send with the revised version, thank the editor and referees for their advice and enclose a list of the substantial changes you have made in response to their suggestions. If you have rejected one or more of the recommendations, explain why.

If you are asked for further experimental information, don't be tempted to rush the work. Do all the control experiments needed, just as you did for the experiments already described.

If the editor offers nothing more than 'further consideration' and the changes suggested are major ones, decide whether the effort of making them is worth while. If you feel that the paper is better as it stands, submit it to another journal after altering it to comply with the second journal's instructions to authors. The second editor may agree with your assessment of the paper's worth.

When a revised article is accepted for publication, and if you find you are likely to be away when proofs are due, tell the editor where and to whom the proofs should be sent, whether to you at a different address or to someone else whom you have asked to check the proofs for you (see (1) above).

(3) *Rejection*

If the editor rejects your paper, read the reasons for rejection carefully. Then decide which of these four steps to take:

(a) If the editor says the article is outside the scope of the journal for whatever reason, send the paper to another journal. Change the style of the article to comply with the instructions of the second journal before you submit it. Some editors use standard letters with sentences such as 'The paper does not lie within the scope of this journal' or 'The journal can publish only about 20 per cent of the manuscripts submitted' even when they really mean they think it is a bad article, so consider whether you need to improve the paper or obtain a few more results before sending it elsewhere.

(b) If the editor says the article is too long and needs changes, consider whether to make the suggested changes – but, again, submit the revised paper to a different journal after changing the style appropriately. The rejection letter would have included an offer to reconsider the article after revision if the editor had been prepared to do this.

(c) If the editor says the referees have found serious flaws in the paper or that the evidence is incomplete, put the paper away until you have obtained more and better information. If you are sure that the editor and referees are wrong, send the paper to another journal or take step (d).

(d) After you have calmed down and thought about it carefully, you may still be convinced that the editor and referees are mistaken in their assessment. If so, write a short but polite letter saying why you think the paper should be reconsidered. Enclose a copy of the manuscript if the editor has returned all the copies you submitted. Do not phone the editor.

IN GENERAL . . .

Don't react too fiercely if you receive adverse comments from the editor or referees. Writing a furious letter or – worse – arguing on the phone won't get your paper published in that particular journal. The criticisms will nearly always have been made for the sake of the journal's reputation and in the cause of science. Even if your critics are mistaken, putting your anger on record won't help. It will be better to cut your losses by considering the recommendations you receive and making constructive use of them.

Referees are often, but not always, anonymous. Their anonymity may annoy you, but don't waste time trying to guess who they are. Your guess is likely to be wrong and you may feel resentful towards the wrong person (or over-grateful to them) for the rest of your career. Editors sometimes put authors and referees in direct contact if this helps to sort out a problem holding up acceptance of a paper.

SUMMARY

(1) Write a brief covering letter to the editor, mentioning any special points about the paper and its contents, and enclosing any releases or copyright forms that are required;   (2) Pack the required number of copies of the paper carefully for posting;   (3) If you hear nothing, enquire about the paper's progress six to eight weeks after the journal received it;   (4) Reply appropriately to the editor when you hear whether the paper has been accepted, will be considered for publication if you make the requested changes, or has been rejected; (5) Remember that editors and referees usually try to make constructive criticisms, and recognize that guessing who anonymous referees are is a waste of time.

CHAPTER ELEVEN

Checking the proofs

Dealing with first proofs   Marking corrections   First reading of first proofs   Second reading of first proofs   Returning corrected proofs and reprint order   Dealing with a second round of proofs

If you ever wonder why it takes so long for manuscripts to get into print, work your way through Figure 4.9. Then count the months it took you to do the work and to write the paper – the publication process is usually shorter, even though many journals have to allow four to six weeks for proof correction.

This chapter applies mainly to journals that are typeset, whether conventionally or from authors' computerscripts. If your target journal is produced directly from authors' manuscripts (camera-ready copy), make as few corrections as possible – but preferably none – in any proofs you receive: all changes at this stage will be expensive and may well be charged to you. If camera-ready copy is used you may not be sent any proofs at all.

When you correct proofs remember that the editor may ignore changes that go against the journal's house style, or stylistic changes that make little difference to the intelligibility of the text. Whatever the production method used by the journal, changes are expensive and, again, journals may charge authors for alterations made in proof, although you won't be asked to pay for typesetter's errors.

FIRST PROOFS

The proofs you receive may be proofs that have already been paged, or they may be galley proofs – long sheets not yet divided into pages. Correct the proofs as described here and as instructed by the journal. Keep changes to a minimum: corrections that change the number of printed lines are tolerable, though not welcome, in galley proofs, and can be expensive and time-consuming in page proofs (see 'Second reading of first proofs' below).

If the journal is typeset directly from authors' disks or magnetic tapes, the proofs you receive may be regular galley or page proofs, or they may be printouts from a laser printer. The output from a laser printer may look slightly different from regular proofs and from the eventual printed page but the correction process is the same as for material typeset in the conventional way, except that you may be asked to correct the disk or tape as well as making changes on the proof.

Sometimes you may be asked to correct an edited copy of your manuscript or printout instead of correcting proofs. Check the edited copy as carefully as if it were a set of proofs. If changes or corrections are needed, mark the manuscript with ink of a contrasting colour to any marking made by the editor or copy editor. If an edited printout includes typesetting codes you may be asked to enter these on your disk or tape. Later, if the publication schedule demands it, the journal staff will check the proofs against the corrected manuscript or a fresh printout instead of sending you proofs, or instead of waiting for your corrected proofs to reach them.

Read proofs twice or more, as described in the sections on the first and second reading of proofs. Mark the proofs as described in 'Marking corrections' below. Deal with them as soon as they reach you: editors usually want proofs back almost before they reach your desk, and missing a deadline may push your paper into a later issue or volume. The editorial office will usually have told you when to expect proofs, so that you can arrange for a colleague to read them if you are away.

MARKING CORRECTIONS

If you have never corrected proofs before, practise on a copy first.

Mark all corrections and instructions in the margins of the proof and put corresponding marks in the text itself. Editors and typesetters won't notice a change in the text if you fail to write the change in the margins.

Use ink of the required colour(s), or of a different colour from any already used on the proofs, and write legibly. Don't make any changes whatsoever on the original manuscript, if this is returned with your proofs: the editor and typesetter will not refer to the manuscript again.

If the editorial office doesn't enclose a list of correction marks you can find them in some dictionaries and style manuals, or use the short list of British and American marks included here (Fig. 11.1). However, your changes will usually be transferred to a master set of proofs in the editorial office or by the typesetter. All you need to do is make your corrections as clear and simple as possible and circle any instructions you put in the margin that don't need to be printed. Typesetters can cope with marks from different proofmarking systems (see below, and next page) provided you follow this advice.

If you are not sure which proof correction mark to use, write an instruction in the margin and circle the instruction. Write the correction or addition beside the instruction and put an appropriate mark in the text to show where the new or corrected material is to go.

The mark in the margin tells the typesetter what the change is. In the British and American systems of proof correction put these marks in the left margin (for preference) or the right margin at the same level as the line in which the change is to be made. For British typesetters put an oblique stroke (a slash) at the end of each marginal mark. If there are two or three changes in one line, arrange them in order from left to right and, for both British and American publications, separate each marginal change from the next one by an oblique stroke. If there are several changes in a line, strike out the whole line and write the new version beside it.

British marks		Instruction	American marks	
Mark in text	Mark in margin		Mark in text	Mark in margin
below characters or words	⟨✓⟩	Restore deleted characters or words	below characters or words	stet
Encircle marks or characters	✕	Remove unwanted marks or reset damaged character, or change font	Encircle marks or characters	✕
⋏	The new matter followed by ⋏	Insert new matter written in margin	⋀	The new matter
⋏	(identified by ⟨A⟩, ⟨B⟩, etc.)	Insert (lengthy) new matter	⋀	The new matter
/ or ⟨2⟩ through characters, or ⊢—⊣ or ⊢══⊣ through words or characters	⟨♪⟩ or ⟨♪⟩	Delete, or delete and close up	/ or — through characters or words	⟨℮⟩ or ⟨℮⟩
/through character or ⊢—⊣ through characters or words	New character(s) or new word(s) (followed by slash mark)	Substitute character(s) or word(s)	/through character(s) or —— through word(s)	The new character(s) or word(s)
under character(s) or word(s)	⟨///⟩	Set in italics	under character(s) or word(s)	ital.
under characters or words	══	Set in capital letters	under characters or words	cap.

British marks		Instruction	American marks	
Mark in text	Mark in margin		Mark in text	Mark in margin
≡ under characters or words	≡	Set in small capital letters	≡ under characters or words	sc
∼∼∼ under characters or words	∼∼∼	Set in bold type	∼∼∼ under characters or words	bf
Encircle characters	⧸	Change cap. to lower case	/ through characters	lc
Encircle characters	⧸	Change small caps. to lower case	/ through characters	lc
Encircle characters or words	⊔	Change italic to roman (upright) type	Encircle characters or words	rom
/ through character, or ⋏	² / ½	Reset or insert as superscript or subscript	/ through character, or ∧	² / ²
⌐	⌐	Begin new paragraph	¶	¶
⌒	⌒	Run paragraphs together	⌒	no ¶
⊔⊓ around characters or words, with numbers if necessary	⊔⊓	Transpose characters or words	∽ around characters or words	tr.
⌐	⌐	Indent	∧	□, or ⊏⊐ or ⧈
⊢⌐	⌐	Cancel indent (move left)	⊏	⊏
⊂ linking the characters	⌒	Close up (delete space)	⊂ linking the characters	⌒
| between characters	Υ	Insert space between characters	| between characters	#
| between characters or words	ⵏ	Equalize space between characters or words	✓ between characters or words	eq #

ᵃBritish marks are based on BS 5261: Part 2: 1976 (BS 5261C:1976 is a useful 8-page card containing the complete list of 67 marks, obtainable from the British Standards Institution at Linford Wood, Milton Keynes, MK14 6LE, UK). American marks are based on ANSI Z39.22-1981.

Figure 11.1 Selected British and American marks for proof correction.

If a correction affects two lines or more, or if there isn't enough space to make changes in the margins, type the corrected version and tape the slip of paper to the left margin beside, but not over, the lines that need to be changed. Don't staple the correction slip or use a pin or a paper clip to attach it to the proof. Identify each slip with an encircled number or, if you are following British Standard BS 5261 (1976), with a capital letter inside a diamond, and add the page or galley and paragraph number. Put the same encircled number or letter-in-a-diamond beside the line of text where the change should go. Instead of using a correction slip you might be able to type the passage near the top of the proof page, if there is enough space – but don't type it at the very top or the very bottom, because typesetters make use of these areas. Wherever you place a typed insert, make it clear where the change in the text is to be made by using the appropriate proof-correction mark in the text, with a corresponding mark in the margin beside the identifying number or letter.

In continental Europe instructions are traditionally written out in plain language, encircled to distinguish them from matter to be included in the text. The proof-correction marks (e.g. those that follow the German standard DIN 16 511:1966) correspond to, but mostly differ in appearance from, the British and American marks. If you use the continental system indicate the place for a change in the text with one of the recognized marks and put the same mark in the margin beside the corrected matter – which can be written wherever there is space for it, not necessarily beside the relevant line of text.

In all systems of proof correction the mark in the text shows where the change written in the margin is to be made. The text mark will usually be an insert mark (a caret), or a line or mark placed beside or drawn through characters or words, or some kind of underlining to indicate italics, bold, or capital letters. Figures 11.2–4 show how some common kinds of corrections are made in the text and margins. Figure 11.5 shows the result.

Remember that the typesetter will make only those changes that are marked in the margins of the proofs. Even if the same errors appear several times, mark *all* the changes throughout the proofs. If you discuss or list changes in a letter to the editor, mark those changes on the proof too (but it is not necessary to list changes in a letter to the editor, unless there are any that need special editorial attention or approval, such as additional material).

The proofs, or the original manuscript returned with the proofs, may have question marks in the margins to draw your attention to passages the editor has some doubt about. There may also be specific queries from the editor or from the typesetter's proofreader. If you want the material to remain as printed, write 'OK as set' in the margin of the proof, circle this message, and put a line through the question mark – but don't erase

141

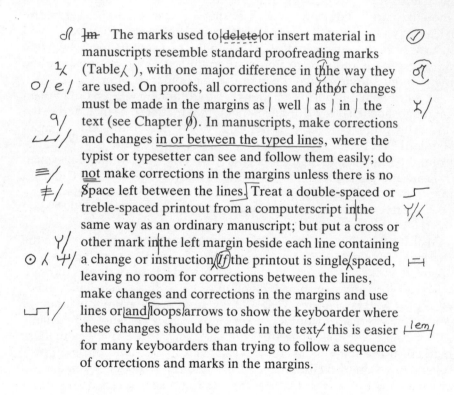

The marks used to delete or insert material in manuscripts resemble standard proofreading marks (Table), with one major difference in the way they are used. On proofs, all corrections and other changes must be made in the margins as | well | as | in | the text (see Chapter ∅). In manuscripts, make corrections and changes in or between the typed lines, where the typist or typesetter can see and follow them easily; do not make corrections in the margins unless there is no space left between the lines. Treat a double-spaced or treble-spaced printout from a computerscript in the same way as an ordinary manuscript; but put a cross or other mark in the left margin beside each line containing a change or instruction. If the printout is single spaced, leaving no room for corrections between the lines, make changes and corrections in the margins and use lines or and loops arrows to show the keyboarder where these changes should be made in the text, this is easier for many keyboarders than trying to follow a sequence of corrections and marks in the margins.

Figure 11.2 Section of proof marked with British proof correction marks (BS 5261:1976).

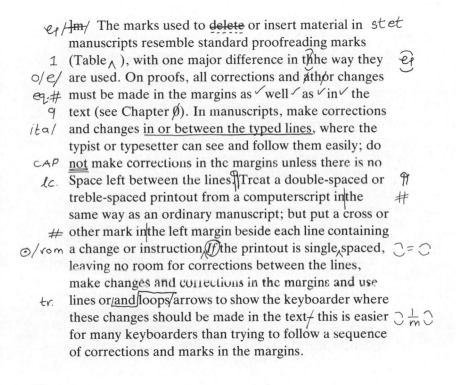

eq /lm/ The marks used to ~~delete~~ or insert material in *st et*
manuscripts resemble standard proofreading marks

1 (Table ∧), with one major difference in the way they *eq*
o/e/ are used. On proofs, all corrections and ~~auth~~or changes
eq.# must be made in the margins as ✓well✓ as ✓in✓ the
9 text (see Chapter ∅). In manuscripts, make corrections
ital/ and changes <u>in or between the typed lines</u>, where the
typist or typesetter can see and follow them easily; do
CAP <u>not</u> make corrections in the margins unless there is no
lc. Space left between the lines¶Treat a double-spaced or ¶
treble-spaced printout from a computerscript in the #
same way as an ordinary manuscript; but put a cross or
other mark in the left margin beside each line containing
⊙/rom a change or instruction. If the printout is single∧spaced, ⊃=⊃
leaving no room for corrections between the lines,
make changes and corrections in the margins and use
tr. lines or/and loops/arrows to show the keyboarder where
these changes should be made in the text⁄ this is easier ⊃ 1/m ⊃
for many keyboarders than trying to follow a sequence
of corrections and marks in the margins.

Figure 11.3 Section of proof marked with American proof correction marks
(ANSI Z39.22:1981).

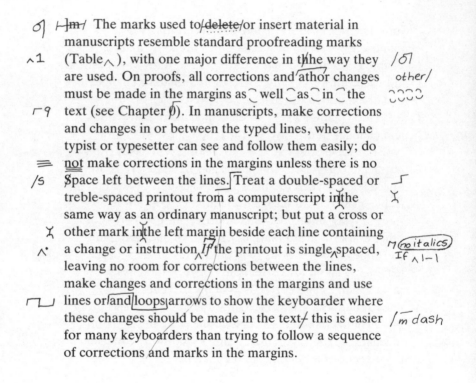

Figure 11.4 Section of proof marked with continental European proof correction marks.

The marks used to delete or insert material in manuscripts resemble standard proofreading marks (Table 1), with one major difference in the way they are used. On proofs, all corrections and other changes must be made in the margins as well as in the text (see Chapter 9). In manuscripts, make corrections and changes *in or between the typed lines*, where the typist or typesetter can see and follow them easily; do NOT make corrections in the margins unless there is no space left between the lines.

Treat a double-spaced or treble-spaced printout from a computerscript in the same way as an ordinary manuscript; but put a cross or other mark in the left margin beside each line containing a change or instruction. If the printout is single-spaced, leaving no room for corrections between the lines, make changes and corrections in the margins and use lines or loops and arrows to show the keyboarder where these changes should be made in the text—this is easier for many keyboarders than trying to follow a sequence of corrections and marks in the margins.

Figure 11.5 Section of proof after correction in the ways shown in Figures 11.2–4.

it. Answer all the questions unambiguously: it won't help if you reply 'Yes' to a question such as 'Should this read X or Y?'

FIRST READING OF FIRST PROOFS

For the first reading of proofs, try to persuade someone familiar with proof correction to read to you from the edited manuscript if this is returned with the proofs, or from your file copy (but the file copy won't show editorial changes). While your helper reads aloud, check the proof for accuracy of numbers, spelling, punctuation, and so on, using a ruler and pen in the way described in the next paragraph. Get your helper to identify new paragraphs, punctuation marks, and words that might be misspelt (for example: 'New paragraph – For the first reading of proofs – comma – try to . . .', and 'Macmillan – that's m, a, c, small m'). Check proper names and unusual words particularly carefully. Make sure that editorial changes have not altered your meaning.

If you can't find a reader, put the manuscript and the proof in front of you. Put a ruler above the first line of the manuscript and place the tip of your pen under the first character on the first line of the proof. Check the proof by looking from manuscript to proof and back again every few words. Slide the ruler down the manuscript page as you check each line, and move the pen along under each character in the proof. Using the ruler and pen like this will help you to focus on the line you are checking and to spot typographical errors. Say the words to yourself as you read; this will help you to catch mistakes such as plural nouns with singular verbs that were overlooked earlier. Take your time.

Check that everything in the manuscript has been printed and that nothing has been repeated. Don't take page numbers, running heads, etc., for granted: check everything. Make sure that the headings and subheadings are correct. If words are broken in wrong or unfortunate places at the ends of lines ('the-rapist', 'off-ending'), mark which characters should be taken back or moved forward. Keep such corrections to a minimum, though, as they may produce even more unfortunate results elsewhere when the paragraph is reset.

Check that the reference list is complete and correct in every detail. Update 'in press' references by adding the volume and page numbers if the papers have been published since you submitted the manuscript. (*Current Contents* is useful for finding recently published papers.)

Make sure that the tables are correctly numbered and that they are as close as possible to where they are first referred to in the text – although the constraints of the page size may mean they are not in the ideal place. If tables have been proofed separately from the text for galley proofs, put a note in the margin – if the typesetter hasn't done so – to show approximately where the tables should appear. Check table contents especially thoroughly.

Make sure that the figures have been correctly made and numbered to correspond to legends with the same number, and that a reference to each figure appears at least once in the text. If the figures are already in position, check that they are the right way up and are reasonably well placed (again, they cannot always go exactly where you would like them to go). If figures have been proofed separately from the text, write the figure number on each, indicate which is the top, and, if necessary, put a note in the margin to show approximately where each figure should be printed.

Unless you receive engravers' proofs, which are themselves of high standard, the quality of the figures will probably not be as good in the proofs as they will be in the printed journal. If you think that the fine detail has disappeared or that the contrast could be improved, say so when you return the proofs. But don't require the printing process to produce illustrations that are better than your originals.

Don't add new material to figures at the proof stage and don't ask for changes to be made unless you find a serious mistake. If corrections are essential, enclose a correct original figure when you return the proofs and identify it clearly as a replacement for the earlier version.

If the journal has redrawn or re-lettered your figures you may receive photocopies of the redrawn or re-lettered versions either before the proofs or with them. Check the redrawn versions particularly carefully.

Check figure legends carefully too, especially if they have been rewritten at a late stage.

SECOND READING OF FIRST PROOFS

Read the proofs again, this time without checking against the original manuscript, to see whether what you have said is accurate and understandable. Make essential changes – but don't start improving the literary style at this stage and don't make minor alterations that should have been made during revision or before you submitted the paper. Don't delete material unnecessarily and don't add new material without consulting the editor. If you have acquired relevant new information that you think is important enough to be included in the article, write an addendum (usually called a 'Note added in proof') to go at the end of the text, before the reference list. When you return the proofs ask the editor's permission to include the addendum and explain the reasons for including it now.

Make alterations in such a way that they produce the fewest possible changes in the lines of text. If you must add essential new wording, try to add it at the end of a paragraph rather than at the beginning – or add a complete new paragraph. Alternatively, try to delete as many characters as you add (characters include spaces as well as letters, numbers, and symbols). If you delete material, try to add the same number of characters as you delete. With modern typesetting methods complete paragraphs can easily be printed out again after a small change has been made but it is still safer – and cheaper – to require as few lines as possible to be corrected.

RETURNING CORRECTED PROOFS AND REPRINT ORDER

If you are sent a duplicate set of proofs, copy your changes and corrections to the second set and keep it. If you receive only one set, keep a photocopy of the corrected set. Put your initials and the date on the first set and return it to the editor or typesetter – the letter or form sent to you with the proofs will tell you which address to use. Post the proof in good time to meet the journal's deadline, using airmail, foreign airmail, or a courier service when necessary. Make sure that the envelope is strong enough and well enough sealed to withstand the journey, that it is clearly marked with the correct address and postal class, and that you have attached a customs form or sticker if one is needed.

If you are ordering offprints or reprints, send your order as instructed by the journal, but don't keep the proofs while you wait for a decision on the size of the order. Post the corrected proofs as soon as you have checked them and say that the offprint or reprint order will arrive later. Then send the order on as soon as possible, especially if you are ordering offprints. Offprints are produced at the same time as the journal and are usually cheaper than reprints, which are produced separately, after the journal has been printed.

SECOND PROOFS

If you are sent a second round of proofs (not common practice for journal articles), check everything against the corrected first proofs to see whether the changes have been made correctly. Check all the lines in the text, tables, or legends in which changes should have been made, and check the lines above and below the corrected lines – corrected lines are sometimes put in the wrong place, or new mistakes may be introduced when the line or its neighbours are reset.

Check whether the tables and figures have been placed as conveniently as possible for readers – but remember that it is not always possible for tables and figures to be printed exactly where they are first mentioned or where you suggested they should go.

Check the running heads or footlines (if any). Examine the title, by-line, and anything else that precedes the text. If you have time, read the whole article through again. Don't, however, make any changes apart from correcting blatant errors. Mark corrections in the same way as for first proofs and return the proofs as soon as possible.

Now all you have to do, apart from getting on with your next piece of work, is wait for publication of the issue containing your masterpiece, and for congratulations from your friends and relations – and perhaps even your colleagues.

Good luck with your next paper too!

SUMMARY

(1) Correct first proofs by reading them twice;   (2) At the first reading try to get someone to read from the edited manuscript while you check the proofs;   (3) Make corrections clearly in the text and in the margins, using ink of a different colour from any already used on the proofs;   (4) During the second reading make sure that the paper is accurate and understandable;   (5) Return the proofs in good time to meet the journal's deadline, and order offprints or reprints if you want them;   (6) If you are sent second proofs, check that the alterations marked on the first proofs have been correctly made.

CHAPTER TWELVE

Preparing short talks and posters

Making effective oral presentations   Content of slides and transparencies   Technical preparation of slides and transparencies
Preparing a short talk   Presenting a short talk   Preparing a
memorable poster

Presenting work at a meeting is an almost obligatory preliminary to
submitting a journal article or a thesis. This chapter therefore discusses
how to prepare short talks, slides for those talks, and posters for
meetings. (For more detailed information on making slides and posters,
see Reynolds & Simmonds 1981, on which much of this chapter is based.
See also Briscoe 1990, Turk & Kirkman 1989, Turk 1985, and Woolsey
1989.)

MAKING EFFECTIVE ORAL PRESENTATIONS

The first presentation you make as a graduate research worker will
probably be to your own department or the institution's journal club, but
you may be giving it before a wider audience, for example during a
meeting of the national society of your discipline. Even if you know
everyone in the audience only too well, prepare your talk carefully: your
career could be at stake. Prepare the paper in good time, rehearse it,
rehearse it again – and yet again, and prepare for the questions the
audience is likely to throw at you.

First find out how long you are expected to speak for. In a ten-minute
talk you'll have time for only about 1000 words (the equivalent of about
four pages of double-spaced typing) and you should include not more
than seven or eight slides or four overhead projector (OHP) transparencies (which take longer to put on than slides). You should therefore
concentrate on getting two or three main ideas over to the audience, not
on sharing every thought you've ever had on the subject with them. In
presentations longer or shorter than ten minutes, increase or reduce the
number of words and slides proportionately.

Secondly, consider who will be in the audience and how much they know about your subject. Match your talk to their level of expertise; never talk down to them. Remember too that listeners, unlike readers, can't turn back to an earlier page when they want to check their understanding of a difficult point.

Thirdly, plan the talk in the same way as a journal article, with an introduction, an experimental section, results, and discussion or conclusions. Before you draft the text decide the main points you want to make and write an outline of the talk (see Ch. 2, 'Constructing outlines'). Then design the slides, as described below.

CONTENT OF SLIDES AND TRANSPARENCIES

Make all slides comprehensible on their own. Don't make them by photographing illustrations prepared for a journal article unless these were designed with slides in mind. Keep the slides simple, omitting as much detail as possible. Avoid abbreviations, apart from those everyone in the audience is sure to know.

Give graphs and tables (if used) a title, placed at the top and separated from the rest of the material by extra space. Use names for different groups shown in graphs or tables, rather than using 1, 2, 3 or A, B, C.

Use your self-imposed quota of slides or transparencies for material that is best demonstrated graphically – don't expect the audience to absorb numerous columns of figures or lines of text. Most of your slides should be bar graphs or photographs whose message can be quickly assimilated. Avoid using three-dimensional bar charts – they may look good but they are difficult to read properly on slides.

If you are making 'word slides', keep to a maximum of 40 words per slide and 40 characters per line. Use no more than 14 lines, including blank lines, per slide. Double-space the lines if there are six or fewer. Keep sentences short.

In a ten-minute talk don't waste a slide on the title of your presentation if it is included in the programme of the meeting or if you know the chairperson will announce titles when introducing speakers. Most listeners find an introductory slide giving the main points of a talk useful, especially at meetings with a large number of papers. In a short presentation you could use your first slide to state the aims of your work and list your main findings.

In the second slide you might show what you studied and, if you used non-standard techniques, how you studied it. Present the information as simply and clearly as possible. You can then devote four or five slides or two or three transparencies to your results. Choose graphs rather than tables (see above). In the final slide state your conclusions and the main points that support them.

In a meeting where everyone is a specialist in your field you might prefer to use all the slides for methods and results.

TECHNICAL PREPARATION OF SLIDES

Make your slides well ahead of the meeting to give yourself time to check, replace, or improve them.

Make lettering large enough to be legible to the naked eye when you hold the slide at normal reading distance. If you can't read the slides under these conditions the audience in most meeting-rooms won't be able to decipher them either. Make sure, too, that the lettering is large enough and bold enough to be read and understood when the slide is out of focus: never trust a projector.

Prepare the artwork – graphs, tables, or wording – within an area of about 200×130 mm (see Ch. 4 for more about preparing figures). Most artwork intended for slides should be about two units high to three units wide (landscape style) (Fig. 12.1), to match the most common format for the projection of 35 mm slides (in contrast, most tables and figures for publication are about three units high to two wide – portrait style). Other slide formats are used but not all can be projected satisfactorily in every meeting-room (Fig. 12.2), so keep to the standard shape whenever you can. Remove any unnecessary spaces between words or columns, and use as few ruled lines (rules) as are necessary for comprehension.

In bar graphs keep to a maximum of six bars. In tables (if used) keep to a maximum of four columns and seven or eight lines (rows), including the title and column headings. Restrict yourself to no more than five or six words on each line of a table that consists of words rather than numbers. See above for the limits on words and characters in word slides.

Write all text horizontally, including labels for vertical axes (the y axis or ordinate) on graphs, so that your listeners don't suffer sore necks trying to read them. Alternatively, read out vertical labels when you show slides containing them.

Use professional lettering methods, such as dry-transfer lettering, stencils, or a lettering machine. Or use a computer graphics program. If you type the text for word slides, use a typeface with capital letters 4–5 mm high (e.g. IBM Directory typeface) within the recommended 200×130 mm drawing area. If you use a smaller typeface use a smaller area for the artwork: 120×80 mm for pica typefaces (10 characters/2.5 cm) and 80×55 mm for elite typefaces (12 characters/2.5 cm).

If your graphs are computer-drawn, try to use a graphics program that automatically produces characters and curves of a legible size and that stops you putting too much text on the slide. Make sure that the printed output is of a high standard; projection will magnify any faults. Curves should be smooth and lettering clear. Symbols must be easy to tell apart (see Ch. 4, 'Lines and curves').

Make curves and rules in graphs and tables 0.35 to 1.00 mm thick (see Ch. 4, 'Lines and curves'). For lettering choose a simple (sans serif) typeface such as Helvetica (Fig. 12.3) or Univers (Helvetica is used in the

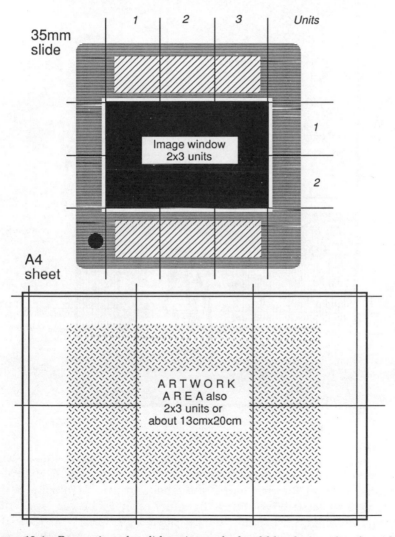

Figure 12.1 Proportions for slides. Artwork should be designed to fit within a rectangle about two units in height:three units in width.

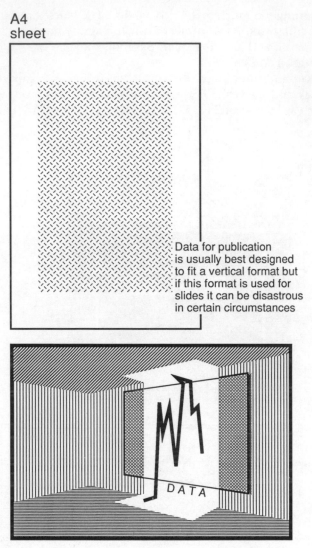

Figure 12.2 Portrait-style slides sometimes spread beyond the screen.

figures in Chapter 4, and Univers is similar). Make capital letters 4–5 mm high, with a line thickness of 0.5 mm – but don't use capitals except for the initial letter of the first word of a label and for the initial letters of words that are usually written with initial capitals. Lower-case (small) lettering, with initial capitals where needed, is easier to read than lettering that is all in capitals. Don't use full stops (periods) at the end of a label, and keep other punctuation marks in labels to a minimum (Evans

1978). Make the space between lower-case characters (letters, numbers, or symbols) about the same as the line thickness of the characters. When you use capital letters, space them individually according to their different shapes. Leave a space of about the same width as the letter 'e' between words. If you are using shading made up of dots or cross-hatched lines, leave spaces of 0.5 mm between 0.5 mm dots and 1 mm between lines 0.35 mm thick (Fig. 12.4). Note that many computer graphics programs produce dot shading which is far too fine. The dots are too close together and reproduce badly. Hatched lines, however, reproduce well if you choose them carefully.

If you are hand-lettering the artwork, start headings and text at the left rather than in the centre: centred lines take more time to position and, because they start in different places, are less easy to read than left-ranged headings and text (Fig. 12.5). Emphasize titles and headings by using bold (heavy) lettering and appropriate spacing (bold type needs extra interline spacing: see Fig. 12.5). Don't underline other words unless it is essential to emphasize them in this way.

Remember that all wording on slides should be in the audience's language, or in one of the official languages if you are attending an

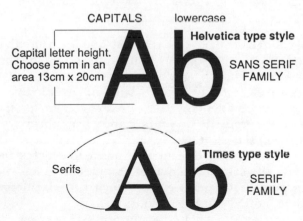

Figure 12.3 Helvetica and Times typefaces.

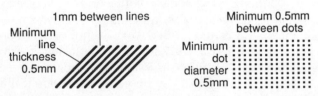

Figure 12.4 Spacing between hatched lines and dots (these examples are not drawn precisely to scale).

THIS TYPE IS ALL IN CAPITALS
AND IS CENTRED.
IT IS MUCH HARDER AND THEREFORE
SLOWER TO READ.
USE CAPITALS FOR ABBREVIATION
OR SHORT TITLES.
AVOID CENTRING TYPE FOR SLIDES

For most European languages the normal reading
direction is from left to right. It is 'normal' to have a left
aligned margin. The use of lowercase lettering makes
for much easier and therefore faster reading. Legibility
can also be enhanced by increasing interline spacing
as in this example.

Interline spacing, sometimes called 'line feed' or
'leading' is even more important when bold
lettering is used. Bold style, as shown here, is
often best for slides especially if white letters on a
blue ground are used.

Figure 12.5 The effects of alignment and spacing on legibility.

international conference, both as a courtesy and to ensure that you will
be understood. Check the artwork carefully, including the spelling.

Use colour where appropriate. You can use it in graphs or tables to get
your message across more effectively – but don't overdo this method of
gaining the audience's attention. Use different colours for specific purposes, choose appropriate colours, and use them consistently
throughout a set of slides.

Coloured artwork is more difficult to produce to a high standard than
black-and-white artwork, but computer graphics programs can simplify
matters. Use pure ('saturated') colours such as red, orange, yellow,
green, and blue, not unsaturated colours such as pink or brown. Use a
maximum of four or five colours if you are using colour as a coding
system, but make sure the colours are as different as possible – blue and
green are easily confused, and some colour-blind viewers may not be
able to distinguish greens from reds, for example. For maximum legibility, again use maximum contrast. (See CBE Scientific Illustration Committee 1988 and Eastman Kodak 1977 for advice on colour for graphics.)

To make your slides as effective as possible, whether they are in colour
or black and white, work closely with your institution's illustration
department (if there is one) and learn as much as possible from what its
members tell you or do for you.

If you are making the slides yourself, put the artwork on a copying stand or easel equipped with 3200 K lights. Photograph it with a 35-mm SLR camera, using daylight slide film with an 80 A blue filter or tungsten film without a filter. Make sure that the artwork fills the camera frame and is aligned with the frame borders. Choose the correct exposure with the help of a test card (grey card). Make extra exposures at one stop above and one stop below the grey card reading. (Heron 1989; see Bishop 1984.)

For maximum legibility most slides are best made as positive images, with black lettering on a white or yellow background. If you use negative images, white on blue is restful for tables or word slides but is less legible than black on white for graphs. Maximum contrast between foreground and background makes for greater legibility.

If you want to produce colour slides from computer images, photo-graph the computer screen in a darkened room. Use a 35-mm camera with a zoom lens and put the camera on a tripod at least three metres from the screen. Compensate for the curvature of the screen if the equipment allows you to do so. A better method is to use a digital film recorder, preferably a high-resolution model. Colours may need some adjustment with this method.

When the film and a spare set of transparencies are ready, put the transparencies in rigid mounts. Glass mounts are preferable to cardboard mounts if the slides are going to be used several times or if they are to be made available to others in your institute. If moisture gets into glass-mounted slides when you travel to a tropical climate, let the slides dry out at room temperature before using them.

Write your name and phone number on a vertical side of each mount. Put a large dot at the bottom left as you read the slide with the naked eye, to show yourself or the projectionist which way up to put the slides in a slide carrier or carousel (Fig. 12.6). Then number the slides in the correct order for your talk, clean them, and test their orientation, order, and legibility in a lecture room. View them from the back of the hall to check that they can be read from there.

Overhead projector transparencies

OHP transparencies are best used when you are talking to a small audi-ence; they are not suitable for use in large lecture rooms. They are easier to make than slides and can be revised at the last minute or even drawn while you are in front of your audience – but it is better to design and make them in good time before your presentation, especially as any shakiness in your hands as you draw will show up all too clearly on the screen.

OHP acetate sheets may measure 250×250 mm or may be the same size as typing paper. Write or draw directly on the sheets, or use a photo-

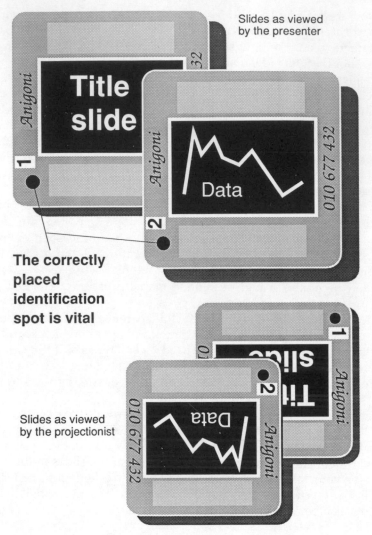

Figure 12.6 Marking slides so that they will be projected the right way up.

copier to copy illustrations onto acetate sheets of the correct thickness (don't use the flimsy kind that melts in photocopiers or laser printers). Don't let your words or pictures go to the very edges of the sheet, though, because the edges may not be visible on the screen. Put the transparencies in cardboard or plastic mounts when they are ready, to make them easier to handle, and write your name and phone number on the mounts. If you don't mount the transparencies put a thin sheet of paper on top of each, to stop them sticking together.

For legibility, make characters on transparencies about 5 mm high;

they will then be about 25 mm high on a 2×2-m screen. Use stencils or dry-transfer lettering in preference to freehand lettering, and restrict yourself to no more than 240 characters, including the spaces between words, on each transparency.

The coloured felt-tipped pens designed for use on OHP transparencies can be either spirit-based for permanent lettering or water-based for temporary lettering, such as the slide number. The pens can have broad, medium, or fine tips. Use strong (highly saturated) colours; pale colours show up faintly, if at all, when projected. Use colour in a logical way, not at random; for example, select one colour for headings and different colours for different kinds of information.

You can either show the whole of a transparency immediately or build up an illustration by adding or removing overlays, or by revealing part of the sheet at a time. Fix the overlays to the frames of the original illustration with tape in such a way that the contents are properly aligned when projected. Using overlays and masks successfully calls for manual dexterity, so practise with them on the projector. Practise using all your transparencies when you practise your talk, and make sure they are clean and legible from a distance.

Other visual aids

Blackboards or whiteboards are provided in many small meeting-rooms, and flipcharts may be available too. If you want to use any of them, check that they will be available and plan how to use them. Amongst other things, they are useful for writing down key words important to your talk when you first mention them. For a non-specialist audience you might also write down your main headings as you reach them, even if you also list these on the first slide.

If movement of some kind is important in illustrating your topic, you may want to show film clips or excerpts from video cassettes. Check whether a film projector or television monitor is available, and allow extra time for the necessary machine to be switched on – and for it to be replaced when it breaks down.

PREPARING THE TEXT OF A TEN-MINUTE TALK

The general rule for presentations is:

Tell the audience what you are going to say;

Say it;

Say what you have said.

One way to prepare such a presentation, after designing your illustrations, is to draft a four-page talk, keeping the sentences short and simple. Remember to pitch the talk at the right level for the audience.

In the introduction (half a double-spaced page) say what question you investigated and why, and what the rest of the talk will contain. Put your work in context by mentioning previous or similar reports, if any. In the experimental (materials and methods) section outline the essentials in not more than one page; do not go into detail about what you observed or used or how you observed or used it. The results section (one and a half to two pages) is the most important part of a short presentation but here again you should include only the most important results. Finally, comment briefly on what your results mean and summarize your conclusions, emphasizing the two or three main points you want the audience to remember.

Put the draft away for a few days before rereading it to see whether it includes all the essential points and omits inessentials. Rewrite the paper as necessary, then reduce it to note form. You may want to write the notes on cards that you can use as reminders during the talk itself. Number the pages or cards in case you drop them before or during the talk.

Another way to prepare a talk is to write notes without writing a preliminary draft.

Whichever way you prepare the text, rehearse your talk aloud, with the slides. Use the notes until you can more or less do without them. Time your presentation or get a friend to do this and to criticize your efforts, including physical mannerisms. Try to do away with unnecessary mannerisms – such as standing on one leg or pacing backwards and forwards – as well as with surplus words. If you can't find anyone to listen to you, tape-record or videotape your rehearsal and put the tape away for a while before re-running it and acting as your own critic.

In addition to rehearsing out loud, you may find it helpful to visualize yourself making the presentation: run through everything in your imagination, from walking up to the platform to hearing the applause at the end and perhaps even to answering questions from the audience.

Make sure that your talk can be presented within the allotted time. You will be forgiven for finishing one minute early but not for going on too long.

Don't plan to read from your manuscript unless this is really the only way you can find the courage to speak. Audiences tend to stop listening, or may even fall asleep, when speakers read their papers or when they spend too much time with their heads bent over their notes. If you have to speak in a language you don't know well, try to persuade a native speaker of the language to make a tape-recording of the talk for you.

Listen to the tape a few times, note the pronunciation of difficult words and the intonation of sentences, and practise the presentation as often as you can.

Because you won't have time to present many details in a short talk you may want to prepare a handout containing additional information about your methods or results. Put the title of your paper at the top of the handout and include your name, the names of your co-authors (if any), and your full institutional address(es). Include the title, date, and place of the meeting in a footnote on the first page. Summarize the talk as well as including the necessary description of your work. Prepare the handout carefully, keeping it as short as possible. Don't put all the copies out at once – keep back a few for people who missed the session.

PRESENTING THE TALK

Before the time for your presentation, check that you have your notes or cards with you. Make sure that your slides are clean and in the right order before you give them to the projectionist, if there is one. Keep a spare set of slides, if possible. Ask the projectionist how you should ask for the next slide ('next slide, please' is the best way if the projectionist can hear you easily). If there is no projectionist, test the remote control, especially the focus button.

If the meeting-room is new to you, look at it before the meeting opens. Check for obstacles on the way to the platform or lectern. Test the microphone, if possible and if there is one, and find out how far away from it you should speak if the audience is to hear you clearly. Find out where the light switches are and who is supposed to operate them. If the screen is large, is there a pointer or torch you can use with your slides? If you are using OHP transparencies, do you know where the switches are on the projector, whether there is a spare lamp, and how to put it in? If you are using a blackboard (or whiteboard) or a flipchart, are chalk and dusters or pens placed where you can find them? Are water and glasses or plastic mugs available?

If the chairperson introduces speakers by making a few remarks about them, try to listen to what is being said when your turn comes. Start by thanking the chairperson in one short sentence for his or her remarks, or by thanking the organizers for inviting you to the meeting, or both, as appropriate. If you can make a light-hearted remark or tell a story before plunging into your talk, do so – but only if you can do this naturally and if the remark or story is 'relevant, brief, witty and dignified' (Booth 1985). If you begin like this the audience has time to adapt to a new speaker and a new topic.

If you are nervous before giving your first talk, don't worry: even experienced speakers feel nervous before a talk. Try to remember that the audience has come to listen to a scientific message and that you are just one of the messengers. Put the message above your nervousness and regard any stage fright you may feel as excitement rather than nerves. Thorough preparation will help you to put your story over successfully. After the first presentation your later ones will become successively easier.

Whether you are nervous or not, don't drink alcohol to relax before your talk. Instead take a few deep breaths to relax yourself just before you start speaking. Make yourself speak SLOWLY, especially at the beginning of your talk, when you will be inclined to rush along too fast for the audience to keep up. Pronounce words clearly, especially key words, for the sake of the 20% or so of your audience who are likely to be slightly deaf (Booth 1985), and speak loudly enough to be heard at the back of the room if there is no microphone. Let your enthusiasm for your subject show through, but don't get carried away by it and start speaking too fast or incoherently. Modulate your voice – don't speak in a monotone. If you are using a microphone, keep it at a constant distance from you. Don't cough into it or scrape chalk against the blackboard. If you expect to cough and there is no water, ask the chairperson for water and continue your talk until it arrives.

Face the audience and look at them, not at the chairperson or the ceiling. Look at each section of the audience in turn. Smile if you can, but avoid phoney friendliness or a fixed grin. Look especially at people at the back of room, but don't fix on a particular person for too long or you will embarrass that person. If you have no choice but to read your paper, look up from the page as often as possible: the more you can remain in eye contact with the audience the more likely they are to listen to you. If you are using notes, don't look down at them any more or any longer than is necessary. Don't pace up and down the platform, jingle coins in your pocket, or otherwise divert the audience's attention. In particular, don't let the audience worry about whether you are going to fall off the platform or trip over cables. Don't smoke.

When you show slides, turn halfway towards the screen rather than turning your back on the audience. Face the audience again after pointing out relevant parts of each slide. If you are using a torch as a pointer, turn it off when you are not using it, to stop the illuminated arrow or dot from dancing around and distracting your listeners. Check the focus of each slide when it first appears and take any necessary action to correct it.

Leave slides on long enough for their contents to be absorbed properly – an average of 14s is enough unless you are using very wordy slides and

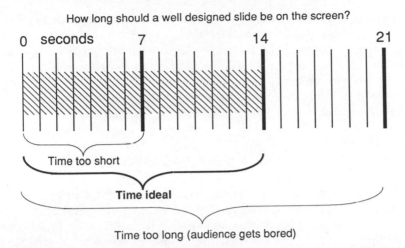

How long should a well designed slide be on the screen?

Figure 12.7 A guide to the length of time a well-designed slide should be left on the screen. Interesting slides can be left on for longer than 14 s but if data are complex it is better to make two or more slides. A final slide can show all the data together, if necessary.

people are taking notes (Fig. 12.7). Flashing through a series of slides at high speed will neither endear you to the audience nor get your message over to them. Don't ask for the lights to be turned on between each slide. If a longish part of your talk doesn't need a slide, include a blank slide for this part rather than calling for the lights or letting the audience be distracted by a slide that doesn't match what you are saying. (Make blank slides of the same background colour as the slides that precede them.) The appearance of a blank slide will also remind you that you have reached a certain point in your presentation.

Some speakers treat OHP transparencies as if they were slides and turn away from the audience unnecessarily. Don't make this mistake: stay facing the audience. Draw their attention to a particular part of an illustration by pointing to the transparency itself, not to the screen. Use a pen or pencil to do this if you haven't got a small pointer. If an illustration isn't needed for a while, put a piece of card over the transparency.

Remember that you must keep to your allotted time for the talk. If the chairperson signals to you that your time is up or nearly up, summarize any remaining material and deliver your conclusion immediately. Leave out slides if necessary (ask the projectionist for the last slide instead of the next slide).

If you are not cut short by the chairperson, make it clear when you have reached the end of the paper. If your words and the tone of your voice don't make it clear that you have finished, you can thank the chairperson, or just say 'Thank you', and stop. If questions are to follow immediately, stay at the lectern. Don't ask for questions yourself – this is the chairperson's job. Take a moment to think before answering questions, and deal with them as calmly as you can. Remain polite even if your questioner seems hostile. Take your notes with you when you return to your seat and remember to collect your slides before the meeting ends (make sure that all of them are still there too).

PREPARING A MEMORABLE POSTER

Instead of a short talk you may be invited to present a poster, which allows personal contact with interested individuals without the formality of a lecture session. Posters are often eligible for publication in congress proceedings but even if your presentation is not going to be published you should prepare it just as carefully as you would prepare a talk.

Content

First outline the content of your presentation and prepare it with your audience in mind. In a typical poster presentation you will be able to include about the same amount of material as in a ten-minute talk: a brief introduction, an outline of the materials and methods you used, a main section devoted to results, and your conclusions. You will also need an informative title that is as short and meaningful as possible, and is – preferably – comprehensible to non-specialists. Leave out unnecessary words: a title such as 'Ultrastabilization of faujasitic zeolite catalysts' is definitely preferable to 'An in-depth study of the structural changes accompanying ultrastabilization of faujasitic zeolite catalysts', while 'How faujasitic zeolites become good petroleum crackers' might be even better (depending on the audience).

Format and design

Find out how posters will be attached to the display board, how much space is available for your presentation, and what shape that space is (Fig. 12.8): does 1×1.5 m mean 1.5 m horizontally or 1.5 m vertically? Are there other restrictions or recommendations, such as the size of lettering for the title, authors' names, and subheadings? When you know the answers to these questions start planning how to fill the space as attractively and informatively as possible. Bear in mind that most people who stop to look at your presentation won't have time to read many words and that a striking picture – provided it sticks to scientific truth – is

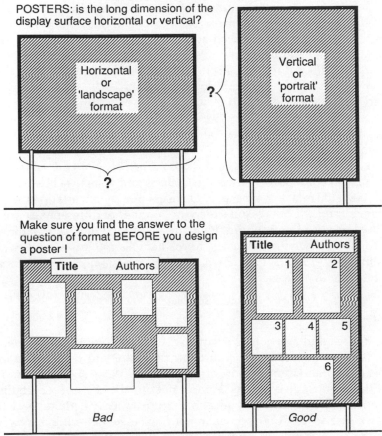

Figure 12.8 Importance of the shape of posters and the way material is laid out.

worth countless words. Plan also how you will cope if the space actually available proves to differ from the amount you were promised.

Using colour is one way of attracting attention. Strongly coloured card makes a good background for text and figures prepared on white or light yellow paper. Photographs and diagrams in appropriate colours will help to catch the eye. You can also use colour for subheadings, for example.

Decide how many illustrations and how much text you can fit into the allotted space. Remember that graphs and photographs are preferable to tables. Start by making a sketch plan of your presentation, drawn to scale. If the space is, for example, about 1.85 m wide by 1.20 m from top to bottom and you want to make your display easy to carry to the meeting, you could plan to fix the pages of text and illustrations to eight pieces of thin card (two rows of four pieces) measuring about 380×540 mm. This leaves room for a title card of about 1.10×0.12 m, large enough for the title to be placed at the top left, followed by your name and institutional address (Reynolds & Simmonds 1981). If you score the back of the title card in the middle you will be able to bend it and carry your whole presentation easily to the meeting. Alternatively, put the title and your name and address on a strip of adding-machine paper which can be rolled up (Day 1988) – but attach the lettering firmly or stencil it directly on the paper, because otherwise parts of the title may drop off.

Each 380×540 mm piece of card can comfortably hold an A4 or 8.5 by 11-inch sheet containing text or illustrations and a smaller illustration of about 150×100 mm (6×4 inches). Devise your own variations on this scheme, but for the best results keep the number of different sizes of text and illustrations to a minimum. Plan an uncluttered display that makes it clear to viewers whether they are to read across the 'rows' or down the 'columns' (Fig. 12.8). A typical reader approaches a poster, stops, reads, understands, and moves on – all in 90 s or even less.

Figures and tables

Make figures and tables in the same way as you would make them for slides (see above) – but use figures in preference to tables, and make figures either landscape or portrait, as needed. Write short but informative titles ('Substance X kills aphids but not bees', not 'Mortality of different insects caused by application of Substance X'). Make these titles the same size as the subheadings in your text (see 'Lettering' below) and place the titles above the figures and tables. Put legends, if needed, below illustrations, and keep them short: nobody will have time to read lengthy legends. Make labels on graphs and diagrams run horizontally, unless they are too long to fit the space available (if they are too long, can you shorten them?). Use photographic copies of illustrations, not the original photograph or drawing, because illustrations can easily be damaged on their way to or from the meeting, or when the display is being taken down.

Lettering

Produce the title and subheadings with a suitable labelling program on a computer, or with a lettering machine, or with dry-transfer or stencilled lettering. A medium-bold sans serif typeface such as Helvetica or Univers is suitable for these parts of the poster (see Fig. 12.3 and Ch. 4 for examples of Helvetica).

Use lower-case (small) letters wherever these are usually used. Use capital letters for words that are usually written in capitals, or for the initial letters of the first word of a label, or for words that usually start with capitals. Do not use capital letters anywhere else: text consisting of capitals alone is much less legible, even at a distance, than an initial capital plus lower-case letters.

Make capital letters in the title at least 40 mm high, large enough to be read from up to 5 m away, with lower-case letters of the appropriate size (about 25 mm for an 'x'). Write your name and address a little larger than the type in the main text if there is space to do so. Subheading capitals should be 10–16 mm high and a lower-case 'x' should be 6–10 mm high. If you are using dry-transfer lettering, the sheets of lettering should contain lower-case letters of the correct size to go with capitals of the size you select.

The text, which has to be readable from a distance of about 1 m, should have capital letters 6–8 mm high, with a lower-case 'x', for example, of 4–5 mm high. If you want to type the text, IBM Directory typeface is large enough, or use a software program that produces high-quality enlarged type. If your typewriter or printer cannot produce text of a suitable size, use dry-transfer or stencilled lettering – which will encourage you to keep text to a minimum.

Final touches

If you plan your display to fit on cards as described here you can fix the text and illustrations to the cards before the meeting or when you arrive at the meeting-place. If you use a glue or paste choose one that doesn't dry too fast, so that you can adjust the position of text and illustrations. To speed up and simplify the layout process on arrival, identify each piece of card with a letter, then number and letter each piece of text and each illustration on the back. Take some double-sided adhesive tape or Blu-Tack, Velcro, or pins with you in case the organizers don't supply these or have run out before you arrive. If the meeting is important, bring or send a complete spare poster: loss and damage do happen, as does vandalism.

Lastly, prepare a handout for interested viewers who want further details. Include the title of the poster and your name and full institutional address. Put the title, date, and place of the meeting in a footnote on the first page. Summarize the poster display and include references to relevant work on the topic. Make the handout as short as possible and make it look at least as good as the poster display. A carelessly prepared handout could ruin the good first impression an interested viewer receives from the poster itself. Write down the names and addresses of people who take a handout. Look up their work and write to them later if your interests coincide.

SUMMARY

(1) Consider the length of the talk and who the audience will be; (2) Write an outline; (3) Design and make seven or eight slides or four transparencies for a ten-minute talk; (4) Draft the text of the talk, concentrating on your results and conclusions; (5) Reduce the draft to notes, rehearse your talk with the slides, and prepare a handout, if necessary; (6) Check the meeting-room facilities; (7) Speak slowly, clearly, and loudly enough to be heard, stay in eye contact with the audience, leave slides on long enough for them to be absorbed, and complete the talk within your allotted time; (8) For a poster presentation, write an outline and consider who the audience will be; (9) Consider the space available and design a legible and uncluttered presentation; (10) Plan how to set the poster up efficiently; (11) Prepare a handout to accompany the presentation.

CHAPTER THIRTEEN

Writing a thesis

Obtaining and reading the relevant regulations   Choosing the
subject, supervisor(s), and title   Planning your work and your
writing   Discussing the structure of the thesis   Keeping records
Planning your reading   Making interim reports and a preliminary
presentation   Choosing the main headings and constructing an
outline   Drafting the sections of the thesis   Typing the thesis
Checking and correcting the typing   Copying and binding the
thesis   Defending the thesis orally

Whether it is for a doctoral, a master's, or a bachelor's degree, a thesis is a
dissertation – a detailed discourse in support of a proposition – in which
you describe your own work or thinking. For the rest of your career in
science you may be working and writing as one of a team. When you
write a thesis you are on your own. The thesis must be all your own
work, although the literature review section will discuss other people's
work. You must prove that you know what you are talking about and
that you can work accurately and think critically. The thesis is your
passport to a career in science or elsewhere and is an important stage in
your life.

Your thesis, sadly, may have a mere five to ten readers: you, the typist
(you again?), your supervisor, the external examiner, and perhaps a few
graduate students who actually make use of what you have written. Your
research is more likely to become known through the papers you publish
in journals. Nevertheless the thesis must be a good solid piece of work,
because your supervisor and the external examiner will be your main
referees for jobs for some time to come. Smart presentation will not
disguise sloppy thinking.

REGULATIONS FOR WRITING A THESIS

The exact organization and appearance of a thesis depend on the
requirements of the academic body to which you are submitting it. Most

169

academic institutions ask for an original document structured somewhat like a journal article. A few expect or allow journal articles to form part or even all of the thesis.

Academic institutions often provide very detailed instructions about typing and binding theses (at least those theses that don't consist of published articles). They rarely give much advice on how theses should be written, apart from general guidance of this kind from the University of Leeds:

> During the examination of your thesis your examiners will be considering both the quality and value of your work and the way in which you have chosen to present your review, results, arguments and conclusions. Your ability to express your findings in a clear and concise manner will be under examination and excessive length or too discursive a style will be judged a weakness.

Obtain the regulations, read them at an early stage, and make a note of the maximum permitted length. Follow the university's regulations and your supervisor's advice in preference to any conflicting advice in this chapter.

SUBJECT, SUPERVISORS, AND TITLE

The subject of your thesis may be suggested by your supervisor or another senior colleague, or you may have found a subject for yourself. Before the title is submitted for approval (if so required) by the university or institute authorities, discuss the topic's suitability for a thesis with your supervisor and at least one other adviser. There must be a reasonable prospect of a successful outcome to your experiments or observations, and the investigations should have a unifying theme rather than being unrelated pieces of work.

For a thesis it is unwise to rely on short-term phenomena such as once-yearly production of larvae, or flowering in the desert: illness or technical failure at the crucial time could set you back a year. You could also lose time if your supervisor becomes seriously ill or moves elsewhere without taking you along. You should therefore find a second, perhaps unofficial, supervisor to whom you can turn if such an emergency arises or if a second opinion is needed at any time. He or she must be an expert who is reasonably familiar with the project.

Choose the title after consultation with your supervisor, remembering that changing the title may be difficult once it has been registered (and see Ch. 5, 'Title').

PLANNING YOUR WORK AND YOUR WRITING

Once the topic has been approved and the title registered, start work as soon as possible and keep working steadily. In the UK, at least, the research councils withhold grants from departments in which too few theses are completed within the required period (usually three to four years for a doctoral thesis). If you want good references it is therefore vital to complete the work and write the thesis before the deadline.

It is fatal to imagine you can easily and quickly complete a thesis after leaving the place where you are doing the work for it, especially if you leave to start a new job. Instead, start writing early – that is, much earlier than you thought you needed to start. One authority says that the best time to start preparing a thesis is not later than two-thirds of the way through the course or time available for your investigation (Booth 1985). You can, for example, write the methods sections even if your results are not complete. If you write the introduction early, be sure to revise it and include the most recent references when you prepare the final version. Beginning even earlier than two-thirds of the way through your investigation will be possible, or essential, if you are preparing a thesis consisting of linked journal articles.

Make a timetable for producing the thesis that gives you interim goals to aim at. If you are still doing experimental work, allow at least six months for writing the first complete draft of a PhD thesis. If you are writing full-time, three to four months should be enough. Allow at least three months for revision. When the final version is ready you should allow one to two months for typing and checking the final version, plus whatever time is needed for binding the finished thesis, undergoing examination on it, and making changes if any are required.

STRUCTURE

Even a short thesis is physically more like a book than like a journal article. A thesis is usually divided into chapters that are more extensive and detailed than the sections of a journal article, the number of tables and figures is not restricted, and at least one copy ends up bound like a book.

Discuss the structure of the thesis with your supervisor when the structure of your project is clear; then choose the section or chapter titles and the main headings within the sections or chapters. A common format is to have a general introduction; a chapter on methods; several chapters each consisting of an introduction, materials and methods (or experimental), results, and discussion sections; and a final chapter consisting of a general discussion and conclusions.

KEEPING RECORDS

As soon as you start your experiments or observations start keeping notes on everything you do or observe. Depending on tradition in your country or discipline you may need to write everything in bound notebooks, numbering the volumes and the pages, and dating the pages when you make entries. If you use notebooks in this way, leave two or three pages empty for a list of contents at the beginning of each notebook but fill up all the other pages in sequence, or draw a line through any space you leave empty between notes on different experiments. Never remove a page from a notebook. To make cross-referencing easier, add your initials and the notebook volume number to the page numbers, or devise some other method for this purpose.

Your notebooks may be the official property of your laboratory or institute. Unless laboratory policy forbids copying you should therefore make photocopies or carbon copies of notebook entries and file them under the different section or chapter headings you choose for the thesis.

Store originals and copies of your notes in watertight fireproof boxes and keep the two sets in separate buildings. Keep duplicate or back-up computer records in a separate building from the originals. Never keep records on a computer only: always print a 'hard copy'. If laboratory policy forbids copying, the laboratory ought to supply a safe for your exclusive use, or give written guarantees of extended funding and deadlines if disaster strikes.

PLANNING YOUR READING

Keep notes and bibliographic details about all pertinent reading. Record the bibliographic information in the comprehensive form recommended in Chapter 6 ('Building a bibliographic database').

Before writing the introduction to your thesis you will have to do a lot of reading or rereading on your chosen topic. To prevent mental indigestion, plan this aspect of your investigation carefully. Start with background reading, then get down to more specialized articles directly relevant to your investigation. Be selective.

Take advice from your supervisor about background reading. You might start with something as general as handbooks and yearbooks, move on to monographs, and then turn to review articles (Ebel et al. 1987). These publications should provide you with further references to more specialized articles. Browsing through current journals in your discipline and using *Current Contents* and specialized information services (for example *ChemInform* or *CA* SELECTS in chemistry) will turn up other potentially useful articles. More references will be produced by a search of other bibliographic databases. If funding is available you might also subscribe to a current awareness scheme that regularly sends you lists of newly published articles matching your 'search profile'. Consult your librarian and the institutional grapevine about the various services available. If your institutional library isn't adequate, try to obtain access to a first-class science library.

Many references turned up by your search may seem worth reading but there will probably be too many to read all of them carefully. If the title and the abstract (sometimes called a summary) are informative and relevant, obtain the full text and look at the tables and figures. If you find these parts of interest, read the rest of the article quickly. Read it more carefully and photocopy it for your files if it answers these questions satisfactorily: Is the problem that led to the work clearly stated? Are the methods used appropriate for answering the question under study and are they clearly described? Are the results presented logically, clearly, precisely, and coherently? Do the results and discussion represent an advance in knowledge of the subject? Another way to read selectively is to go straight from the title to the results section and its tables and figures; again, read the rest of the article only if the tables and figures are well presented and the results sufficiently interesting and relevant. (See Haynes et al. 1986 and other articles cited by Huguier et al. 1990, p. 133–136.)

EARLY STAGES:
INTERIM REPORTS AND PRELIMINARY PRESENTATION

Long before you start writing the thesis you will probably have to write one or more interim reports for your supervisor, co-author a journal article, present a paper or poster (see Ch. 12), or even undergo an oral examination on your work to date. These exercises in communication should be welcomed. You will gain some practice in writing and will probably be able to weave some of the material, especially the materials and methods (or experimental) sections, into the thesis. Feedback from a seminar presentation may point to alternative interpretations or new lines of work.

When you write an interim report start by choosing the main headings and making an outline (Ch. 2, p. 13 et seq.). You will probably be expected to concentrate on the methods and results sections, including tables and figures, but you should include a title, an informative abstract, and a short introduction and discussion. Mention also any conclusions you have reached and indicate what you intend to do next.

173

If you write or co-author a journal article before writing your thesis, follow the journal's instructions to authors and the advice in Chapters 1–11. If you use material from the article in your thesis, or if the article forms part of the thesis, make it clear which parts of the work were done by you and which parts by your colleagues.

Design tables and figures as your work proceeds but don't make the final versions until you revise the draft of the text (see Ch. 3 & 4).

CHOOSING THE MAIN HEADINGS AND CONSTRUCTING AN OUTLINE

Start your thesis by choosing the main headings and constructing an outline, as described in Chapter 2 (p. 13 et seq.). When you make the outline, remember that you will be able to write at greater length in a thesis than in a journal article; in particular you will be allowed to speculate more in the discussion than is usual in journal articles. Nevertheless don't plan to be prolix: aim for less than half the maximum permitted length. The examiners will be delighted to see a short thesis, provided it expounds your theme clearly and demonstrates your research abilities sufficiently well – consult your supervisor if you have any doubts about the length. In general, don't omit experiments that failed: the examiner needs to assess how you used your time. If some experiments failed, say so, say why, and say what you learnt or would do differently next time; that is, show that you can think critically. Your supervisor will advise you on how best to do this. Every scientist has reverses, although these are seldom revealed in print.

One of the worst calamities you can run into arises if your work shows beyond reasonable doubt that a long-cherished dogma is fallacious. This is especially catastrophic if your supervisor or external examiner is the dogma's originator. A situation like this requires a rare combination of tact and fortitude. Start by talking to your supervisor. Present your problem openly and honestly and base your next steps on that interview. Never agree to publish anything you can't defend wholeheartedly.

DRAFTING THE SECTIONS OF THE THESIS

When you get down to the first draft, follow the advice in Chapter 5 on practical preparations for writing and other matters. Write the sections of the thesis in the same way as you would write the sections of a journal article (Ch. 5), with any differences referred to below (and with the difference that you must usually write a thesis title at an early stage). You

. . . your work shows beyond reasonable doubt that a long-cherished dogma is fallacious.

will probably find it easiest to write the materials and methods section(s) first, then the results. Write the abstract when the rest of the draft is complete. Headings are useful signposts for readers (see Ch. 2, 'Deciding the structure: choosing the main headings'), so use subheadings for subsections within each chapter. And because theses rarely have indexes you should also plan to include 'running heads' on each page when the thesis is typed (see 'Typing the thesis' later in this chapter).

Introduction

In the introduction you are expected to describe the background to your investigation. Just as in a journal article, you must answer the questions who?, what?, when?, where?, and why? Say why you did the work and what its purpose was. Announce the general topic, say which particular aspect you are dealing with, state the question(s) you intend to answer, and explain the importance of the work. Then set the scene in some detail, in the way that review articles do (see Ch. 14, 'Writing a review article'). That is, describe and comment briefly on recent work on the topic. Do not, however, list all the papers on the subject since the days of Galen or Linnaeus . . . unless the regulations so require. Identify the most important contributions in the field. If your study is the first you know of on the topic, survey the articles or other publications on related topics that led you to decide on this particular study. End the introduction with a sentence that leads into the next section or chapter, such as:

It therefore seemed appropriate to try to discover how much of effect Y was due to the presence of substance Z.

Materials and Methods (or Experimental)

Write the materials and methods sections in the way described in Chapter 5, with full details of what you studied and how you studied it. Provide enough information to allow other scientists to repeat the work and verify your results. Your readers, other than your supervisor and examiner(s), will probably be graduate students, so make your explanations as helpful as possible.

Results

Write the results sections as described in Chapter 5. Present your data in tables or figures, as necessary. Organize your material as clearly as possible – not necessarily in the order in which you did the experiments. Keep the examiners and other readers in mind: they will want to understand your findings easily.

Make the final versions of your tables and figures at the revision stage (see Ch. 3 & 4). At that time be sure to make enough copies to fulfil the institution's requirements. Avoid tables that stretch over several pages; instead, divide a large table into tables of one or two pages each, or put large tables in the appendix and provide a summary table or graph in the body of the thesis. The summary table should show, for example, average values and results of statistical comparisons. Such summaries are easy to grasp and should be used whenever possible.

Figures, too, are better kept within the limits of the size of typing paper you will be using. If this is difficult and a figure cannot be made into two smaller figures, make a large figure that can be folded, in preference to reducing the figure so much that it becomes unreadable – even though a folded figure will produce problems when the thesis is photocopied or microfilmed.

If possible, make all tables and figures 'portrait' style (height greater than width) to save readers from having to turn the thesis sideways. Remember to leave a margin of the required size on the left, to allow for binding. If the final version is to use single-sided pages and you want to put legends facing the figures, put the wide margin on the right on pages with legends – and remember that you will have to rearrange these pages for photocopying and again before binding, because otherwise the blank side may be photocopied instead of the side with the legend.

Summarize the results in a set of concise, numbered statements.

Discussion

Write the discussion in the same way as the discussion section of a journal article (see Ch. 5): that is, examine your results, explain their significance, and answer the question(s) you posed in the introduction. You can allow yourself to say more than you would in a journal article but don't waste words, here or anywhere else in the thesis. Stick to the subject – your results – and don't be tempted to wander off into another review of the literature. You must, however, set your findings in the context of the existing body of knowledge.

If you are contemplating combining the results and discussion sections, don't. Examiners find separate sections easier, and combining them can seem slipshod.

Conclusions

A chapter entitled conclusions may not be obligatory but can be very helpful to readers (especially examiners). State the significance of your results briefly. Don't claim too much, and don't hedge your claims unduly (see Ch. 8, 'Verbosity and pomposity').

References

If the regulations stipulate a particular style for the reference list, follow that style carefully. If no style is laid down, provide an alphabetical list of references (Harvard style), without numbers (see Ch. 6, 'Styling citations and the reference list').

Cross-check the references, making sure that every citation in the text is included in the reference list and that every reference in the list is cited in the text. Make sure that all details such as volume and page numbers are included and are correct, and that the list is in the correct order.

Appendix

If you want to include lengthy material related to your theme but not essential to the development of your argument, put that material in an appendix (regulations permitting). If you cite published work in the appendix, add the references to your reference list.

Notes and footnotes

Keep endnotes (collected at the end of the chapter or thesis) and footnotes (typed at the bottom of the page) to a minimum, as they interrupt the reader's concentration on your line of argument and tend to look like afterthoughts. Follow university regulations about the way you set out the notes and footnotes. Don't use miniprint when you type the notes.

Acknowledgements

Include an acknowledgements section, mentioning all the help you have received. This section is often placed before the introduction.

Abstract

When the rest of the thesis has reached the second or third draft, write the abstract as described in Chapter 5. Keep the abstract within the limits laid down by the university.

Remember that the title and abstract may appear separately in a thesis-abstracting publication and must therefore be comprehensible on their own. Include as many significant words from the text as possible, to ensure that the thesis can be tracked down in a computer search.

TYPING THE THESIS

Have your thesis typed on a word processor, if possible, as this makes revision quick, easy, and cheap. If you can type reasonably well, obtain the use of a computer, preferably one with plenty of memory and a hard disk, and type at least the drafts yourself. Even if you have to cope with an unfamiliar word processor, typing your own drafts will probably be quicker than handwriting a version that a typist unfamiliar with scientific terminology can read. Remember to back up your files regularly. Make sure that your files will print out correctly on the printer you intend to use. If you plan to have the final version professionally prepared from your disks, make sure that the disks and your word-processing program are compatible with those used by local typing or printing services.

Before you start preparing the final version, reread the university regulations. Size and weight of paper, typeface, margins, and space between the lines of text are among the points to note before typing starts. Note too whether single-sided or double-sided text is preferred, whether pages should be numbered consecutively throughout the thesis, including tables and figures, and where page numbers are to be placed. See also what information must be included on the title page, whether a signed declaration about the work must be inserted, and in what order the various parts are to be presented. Make a checklist of these and other relevant points for yourself or whoever types the thesis for you.

Use opaque typing paper of good enough quality to ensure a long shelf-life for your thesis, and change the typewriter or printer ribbon immediately it starts becoming faint. Make sure that the margins are wide enough for binding. For single-sided text leave a left-hand margin of not less than 40 mm (about 1.5 inches) and make the other three margins not less than 20 mm wide or deep – or whatever size the regulations specify. If you are allowed to use both sides of the paper, make both left and right margins 40 mm wide (or as specified) on all pages, to allow for binding.

In typing the text, follow the advice in Chapter 9 unless it conflicts with university regulations or applies only to material that will be typeset. Type the title page in the form required by the university, and prepare the table of contents last of all, when the manuscript pages have all been numbered. On each page type a running head consisting of the main section heading that applies to the page; start this headline at the left margin and leave an extra double line space to separate it from the text below (some word processors can change the running heads automatically to match headings in the text).

CHECKING AND CORRECTING THE TYPING

Check the final manuscript carefully. Get someone else to check it too, if possible. If you can't spell, find a helper who can. Use a spelling-checker program if it is supplied with your word processor but don't rely on it to remove all typing errors or expect it to eliminate grammatical errors (not even style-checking programs will produce perfect grammar for you: see Ch. 8, 'Problems of grammar and style'). Check all numbers and proper names particularly carefully. Make sure that tables and figures are correctly numbered and referred to in numerical order in the text (don't refer to Table 5 before Table 4 without good reason). Check the references carefully too.

Correct minor errors with correcting fluid, using as little as possible, and write or type the corrections in when the fluid is completely dry. If there are several corrections to be made on a page, type that page again. However, if all copies of the thesis can be photocopies and you are typing the thesis yourself, you can make corrections on adhesive correction tape. Carefully made alterations will be invisible on the photocopies.

COPYING AND BINDING THE THESIS

Before you make copies of the thesis, arrange it in the required order. If the figures are being photocopied, put them (and the tables) as close as possible to where they are referred to in the text – unless the university asks for them to be placed elsewhere.

When the thesis has been copied, check every copy to ensure that no pages are blank, upside down, illegible, missing, or out of sequence. Put all the copies on a large table and compare them page by page. Make sure that the 'top' copy (the original) is not mixed in with the duplicates. Check the photocopying of photographs and other figures particularly carefully. The bindery will not look after any of these things. Note also that the pages will be trimmed in the bindery, so make sure that no material intrudes into the margins.

Before you have the copies bound ask your supervisor whether the examiner will agree to read an unbound copy. The task of rewriting sections or making major corrections will be much easier if the examiner accepts this suggestion; you will also save on binding costs.

When all the copies are correct and complete, arrange for the necessary number to be bound and for the required lettering to be placed on the front covers and spines. Check the bound copies again in case any pages got lost or misplaced during binding. Then submit the copies as required and get a receipt for them.

DEFENDING YOUR THESIS ORALLY

The oral examination for a thesis may be anything from a simple formality – a brief discussion with the examiner and your supervisor – to a public examination of your knowledge of your field of science, not simply of the topic of your thesis. Unless you are sure the examination will be of the first kind, prepare yourself for it thoroughly. Reread the thesis carefully, especially the parts you wrote earliest. Bring yourself up to date with recent work published on the subject – which you may not have had time to read while writing the thesis. Try to foresee what the examiners will ask you (Ebel et al. 1987, p. 54). Prepare for questions about why you chose one method or one course of action rather than another, why you interpreted results one way rather than another, what you would have done if the results had been different, and what direction future work on the same topic might take. If you are likely to be asked to summarize your thesis in 10–15 minutes, prepare a summary that you can present without a manuscript or notes. When the time comes, listen to the questions carefully – don't answer the hypothetical questions you put to yourself instead of those actually put by the examiners. Breathe deeply and stay calm: on this occasion you are the expert who knows most about the work being discussed. It is highly unlikely that you will be failed at this stage.

SUMMARY

(1) Read the regulations for the presentation of theses;   (2) Consider the subject, structure, title, and timing of your thesis and who your supervisor will be;   (3) Plan note-taking and reading strategies;   (4) Make a preliminary presentation or write interim reports or journal articles;   (5) Design tables and figures;   (6) Write an outline, then draft the rest of the thesis;   (7) Type and correct the final version;   (8) Obtain a receipt when you submit the bound copies of the thesis;   (9) If you have to defend your thesis orally, reread it and prepare carefully for the questions likely to be asked.

CHAPTER FOURTEEN

Writing review articles and book reviews

Types of review article   Searching the literature   Writing a review article   Writing a book review

When you are established in your career, writing review articles and book reviews will provide a way of sharing your ideas about the development of your field of science. Writing reviews of these kinds will make you adopt a wider perspective than is usual in a research paper, so helping your own development.

TYPES OF REVIEW ARTICLE

Review articles are more or less detailed summaries of the work published on a specific topic. They either evaluate published work on the topic and provide an up-to-date synthesis of it or summarize the work without evaluating it. In clinical medicine meta-analyses are reviews that pool and summarize results from reports of several clinical trials dealing with the same question, or thoroughly evaluate the methods used in those trials, or both summarize and evaluate the reports. Meta-analyses are also used in other disciplines.

A good review, like other research projects, starts with a well-defined question and either answers that question by a systematic search of the literature or produces new questions for further research. Unpublished work by the author should not be included in a review.

Reviews are usually commissioned by editors – who sometimes prefer young and hungry scientists to the more famous who are often too heavily committed to keep to a deadline or find time for the necessary reading. And instead of asking for page charges, as some journals do for other kinds of articles, editors may even offer you a small fee for a review article.

Before you agree to write a review, look at the journal's instructions to authors and at reviews in the current issue to see what kind of review you are expected to produce. The editor's letter of invitation may define the exact scope of the required article. If not, ask for a definition and discuss whether unpublished material or material that is not easily available ('grey literature') should be cited. Such material may be extensive.

Topics for review articles, including meta-analyses, may also be proposed by authors who have noticed a gap in the literature. If you have such an idea, discuss it with an editor before getting down to serious work. Describe the scope of the intended review and list the subtopics you plan to cover. Make it clear that you know what kind of review the journal usually publishes and that you plan to match the journal's requirements. Give the editor a realistic estimate of when your review will be ready.

Wherever the idea for the article originates, you must be clear whether you are writing a critical review assessing a selection of published work or providing an annotated list of all the relevant work published during a given period.

SEARCHING THE LITERATURE

You will obviously have to do a lot of reading before starting to write a review article. When the scope of the article has been agreed with the editor, decide the main headings under which you will tackle the subject (see below). Then make an outline of the article (see Ch. 2, 'Constructing outlines') and plan your reading strategy (see p. 172 for comments on reading selectively). Start with background publications, such as textbooks and earlier general reviews, before turning to the specialized work that is your subject. Make notes on your procedures, on the selections you make, on the criteria you use, and on everything you read. Keep the notes and bibliographic details on filing cards, looseleaf pages, or in a computer. Don't forget that you have to report quantitatively on all the references published during a given period and on the references you comment on in detail in the review. Guard the notes and bibliographic information carefully – forestall disaster by keeping a second set somewhere other than your regular place of work. Make photocopies or back-up copies of new or altered material at regular intervals.

As you read, your ideas on the topic will almost certainly change and broaden. Your reading may then have to be extended beyond the original limits, and the original question may have to be widened or narrowed.

As you read, your ideas on the topic will almost certainly change and broaden.

WRITING A REVIEW ARTICLE

Review articles, like other articles, should be written with their intended readership in mind. Even if you are writing for a specialist journal, some readers will be non-specialists needing an overview. Orientate such readers by providing a clear and simple introduction and by stating your conclusions clearly. If you are writing for a general journal the readers are even more likely to need a comprehensible and comprehensive survey.

Review articles do not usually have the standard headings – materials and methods, results, discussion – used in many research articles. After the introduction, they commonly continue with a description of the literature search and end with conclusions and recommendations for future work. The headings in the middle of this sandwich (equivalent to the results section) depend on which aspects of the main topic are being discussed. A review on chemical change in the earth's mantle might, for example, include these headings: composition of the mantle, chemical variations, petrographical evidence, volcanological evidence.

In the introduction describe the background as clearly as possible and state the question you set out to examine. In the literature search (or scope and methods) section say whether you used conventional or computer-based searching methods, or both. Name the bibliographic databases you screened and the earliest and latest dates covered. Describe your criteria for choosing key words for the search and for selecting the articles to be assessed or annotated in the review. If the review is a meta-analysis, say how you assessed the reliability of the studies covered and their publication bias (the tendency to exclude negative studies). Explain the statistical and other methods you used to analyse the literature.

In the main part of the text summarize or evaluate the selected articles as objectively as possible. Indicate the limitations of the articles. Analyse variations in the findings critically. Set out your conclusions and rec-ommendations for further work clearly. Don't hide treatment of a controversy under a bland heading. Present both sides of a controversy as dispassionately as you can, and state your position.

In your conclusions summarize your main findings and recommend directions for new research.

Make the reference list – likely to be a long one – as accurate as possible, always taking the details from the original publication. Reference lists in review articles are a valuable source for many researchers; mistakes are unprofessional and will waste a lot of people's time.

Follow the journal's recommended style for references. Check that each citation in the text is linked to a (correct) entry in the reference list and that every entry in the reference list has a corresponding (and correct) citation in the text (see p. 85). If you want to include extra references that are not referred to in the text, put them in a separate list of 'Further reading'.

Include an informative-indicative abstract that is as informative as possible (see Ch. 5, 'Abstract'). State the purpose of the review, say what search methods you used, summarize your main findings, and indicate your main conclusions and recommendations (Squires 1989).

WRITING A BOOK REVIEW

Like review articles, book reviews can be either evaluative or descriptive, although most are evaluative. Occasionally book reviews consist of essays on the subject covered by the book, with perhaps only a brief mention of the publication itself. The essay type of review is usually used only in journals that specialize in book reviews. Only the evaluative type of review is discussed here.

Book reviews are nearly always written at the request of a book review editor. If you are interested in reviewing a particular book, contact an editor first, to make sure that your review will find a home and to explain your interest in the book (if you are a close friend or a sworn enemy of the author the editor will need a less biased reviewer).

Specialist journals rarely pay book reviewers, although general jour-nals may pay a small amount per 100 words. Your reward for reviewing will most often be the review copy and the pleasure of seeing your name on the published review.

If you are asked to write a book review, make sure that you can submit it by the required date – if not, either return the book or negotiate a new deadline with the book review editor. Book reviews in scientific journals tend to appear several months after the book's publication and this time lag should not be made any longer than is necessary. Textbook reviews, in particular, need to be published well before course books for the coming year are chosen by lecturers and ordered by college bookshops.

Before you start work on your review, read several recent book reviews in the journal. If the journal sends you a set of guidelines on book reviewing, observe the technical and intellectual requirements set out there. The guidelines or the editor's letter should state the approximate length of the review and you will probably be asked to type it in double spacing, on one side of the paper. The guidelines may point out that the journal requires fair and balanced reviews, written in temperate language – that is, you mustn't attack the authors or editors of the book personally or make libellous statements.

You should read the whole book unless it is a dictionary or a fat textbook, when you can make spot checks or read sections at random instead. Read the preface or introduction, or both, and read the blurb – often written by the author, then sometimes jazzed up by the publisher's marketing department – to see what the author or editor claims the book sets out to do and who it is for.

In your review describe the contents of the book briefly and say which readers will benefit from it, especially if your view of the readership doesn't coincide with the publisher's claims. You may want to start by sketching in the background of the subject and mentioning the author's previous work or reputation. Then evaluate the book on its own terms, saying how well you think the author or editor has achieved the stated aims. If necessary, criticize those aims and the way the subject has been handled. Be constructive in your criticism. Compare the book with other recent books on the same subject, if there are any. If it is a new edition, especially of a textbook, say whether the changes are substantial; the secondhand market is important to students and a new edition can kill the market for an earlier edition.

Take the following points into account in your review, as well as any others that come to mind:

Is the material well selected, well organized, and up to date?   Are the arguments clear and logically correct?   Are the statements of fact accurate?   Are the conclusions convincing, original and important for the discipline as a whole, or for the special topic of the book?

Where experimental work is discussed, is the experimental design satisfactory? Is the style clear, concise, and readable? If the book has several authors, does it have sufficient unity of style and content?

You may also want to comment briefly on some or all of the following:

The general appearance of the book   The legibility of the typeface and (if it is a paperback) the durability of the binding   The incidence of typographical errors (but only do this if the number seems excessive)   The speed of publication if the book is a symposium or conference report   The clarity of the illustrations and their legends The accuracy and coverage of the references, and whether they are up to date   The usefulness of the index, or its absence.

There is no need to comment on the price of the book unless it is particularly expensive – or cheap. Readers can decide for themselves, with the help of your review, whether the book is worth the price. Resist, too, the temptation to tell the publishers they should have produced a paperback to keep the price down. Paperbacks cost nearly as much to produce as hardbacks and are only sold more cheaply because the publisher expects to sell a large number of copies and, usually, to cover most of the initial costs with sales of the hardback.

The book review editor may say that you will be sent a copy of the journal issue in which your review is published. If not, and if the journal is one you don't usually read, ask the editor to send you a copy of the issue if you want one for your collection of publications.

SUMMARY

(1) Consider whether you will write a descriptive or evaluative review article, or a meta-analysis;   (2) Plan a systematic reading strategy;   (3) Write a well-structured article with the readers in mind;   (4) Consider what kind of book review you are going to write;   (5) Read the book(s) you are reviewing and write a constructive review.

CHAPTER FIFTEEN

Preparing successful grant proposals and curricula vitae

Choosing a project   Choosing a likely funding body   Obtaining instructions and application forms   Contents of a proposal   Drafting the title, abstract, and research plan   Drafting the budget Assembling a coherent proposal   Preparing a curriculum vitae for a grant proposal   Preparing a curriculum vitae for a job application

Applying for a grant to start or continue research has a lot in common with writing a journal article and is a chore most researchers have to cope with many times in their careers. This short chapter outlines a general approach to the preparation of grant applications but doesn't go into the detailed requirements of particular funding bodies. It also discusses curricula vitae, which are an essential part of every application for a grant or job.

GRANT PROPOSALS

Choosing a project

The first and most important requirement for a grant proposal is to have a project likely to attract funding. Just as for a journal article, you need to formulate a question worth answering. You must also suggest how the problem leading to the question can be tackled in a way likely to produce a solution. Your proposal must be coherent, not a ragbag of ideas on different topics. And when you make your first request for funds for a specific project it will help if you have already done some pilot work on the problem. Such initiative will show the assessors who review your proposal that you are a self-starter likely to succeed with a project, whether it lasts three months or three or more years.

Choosing a likely funding body

When you have a suitable project in mind, look around among the possible funding bodies. These bodies range from international agencies to small private foundations and perhaps even your own research organization. Senior members of your department will be able to suggest potential targets outside the organization; libraries should be able to help too. Don't waste your time obtaining information from unsuitable agencies or from the wrong section of a large funding body. A funding body specializing in grants for major programmes costing millions is not ideal if your project is a minor one costing a few thousand, nor is one specializing in engineering ideal if you are a botanist.

Obtaining instructions and application forms

When you have chosen one or more targets, obtain and read the instructions or guidelines for applicants and the necessary application forms. Some funding bodies take bureaucracy to extremes and return applications that fail to observe trivial technical requirements. You should therefore read all the information carefully and draw up a checklist of everything that has to be included in the proposal. If any colleagues have applied successfully to the same funding body, borrow copies of their proposals to get an idea of what that agency likes to receive.

Before preparing a grant proposal make sure that your project meets accepted ethical standards (see p. 2). Consider too whether publication of a report or journal article about the project will break any official secrecy regulations or invalidate a later application for a patent.

Proposals take time to grind their way through the selection machinery. The funding body's guidelines usually indicate how far ahead of your planned starting date for the project you should apply. Allow two to three months for preparing the proposal and start work on it at least six months before the project's starting date – or up to 12–18 months beforehand for the larger agencies, even if the appropriate committee meets every month. Quick decisions may, however, be possible if your application is for a small amount (in 1988 the Medical Research Council in the UK classified amounts under £20 000 as small).

Note that some funding bodies ban applications that are being considered by another organization, just as journals ban duplicate or multiple submissions. Different agencies may use the same reviewers and applicants can easily be caught out if they break the rules. If the funding body to which you are applying allows duplicate applications, say whether an application for the same or a related project has been sent to another agency.

Some funding bodies encourage would-be applicants to talk about projects with their staff before submitting a full application. Other bodies may ask for a written outline first. One such agency recommends, for example, that applicants should submit two or three single-spaced pages describing the proposed research and including a brief curriculum vitae, a list of publications, and an approximate costing of the project. If the preliminary outline is satisfactory the applicant is then asked to submit a full proposal on the appropriate forms.

Although contact with a funding body's staff may be encouraged, attempts to influence the scientists responsible for assessing a grant proposal are frowned on. Nevertheless – just as for a journal article – you need to know who your readers will be, so find out whatever you can about named members of relevant committees. If you know any such assessors, it can do no harm to listen to anything they say about the kind of topic or approach the funding body currently favours – but tell them openly that you intend to apply for funding.

Contents of a grant proposal

Grant proposals are likely to include some or all of the following:

Table of contents   Abstract of proposed research   Progress report on either a previous proposal or the preliminary findings that led to the current proposal   Description of research plan   Introduction: purpose and background of current proposal   Specific aims to be accomplished   Methods and materials (or human or animal subjects) to be used   Results expected   Discussion of significance of project   Reference list   Budget and budget justification   Details of publications arising from previously supported projects Reprints of relevant publications   Details of current grants held by the applicant   Curricula vitae of applicant, collaborating research workers, research assistants   Signatures from departmental heads or university authorities supporting or approving the research plan and budget.

The list above is not exhaustive, and the order and amount of detail required will of course vary from one agency to another. You may even get away with sending in just a brief outline if the agency is both small and enlightened. The guidelines for applicants usually tell you how to arrange your proposal, what headings to use, how long each part should be, and so on. If there are no guidelines or application forms you can arrange the application as you wish – but remember that the assessors will judge you on the organization of the proposal as well as on the brilliance of the ideas it contains.

Drafting the title, abstract, and research plan

The title, abstract, research plan, and budget are the core of any grant proposal. Prepare the first three in the same way as you would prepare a journal article (see Ch. 2–5). That is, write down the main headings and the points that belong under each of those headings, construct an outline, design any tables and figures that are essential, and keep your readers and the funding body's guidelines always in mind as you draft the text.

Some of the assessors who review the final proposal will be experts in your topic but others may not be: keep both categories in mind when you are writing. For most grant proposals you should write as clearly, concisely, and simply as possible. Some funding bodies, however, favour proposals written in Eurospeak or sociologese. If you are applying to one of these, employ an editor/translator or study the language of earlier successful proposals carefully.

Write a title that keeps within the length stipulated by the funding body and that is accurate, concise, informative, and preferably memorable as well as easy to understand. The title will be used to refer to your application many times during its journey through the reviewing system, so make every word earn its place.

The abstract – or summary, or synopsis – must also keep to the required length and be accurate, concise, informative, and easy to understand. The abstract is usually the first part of the proposal that assessors look at, after the title, and it may be the only part some of them ever read. If no length is specified, try to limit the abstract to not more than half a dozen short sentences.

The abstract will help assessors to remember what your proposal is about when they are writing their reports. Say briefly why you think the project should be undertaken, what you plan to do and how you plan to do it, what you expect the results will be, and what the significance of those results might be.

A typical description of a research plan includes the same parts as a journal article, usually in a shortened form. In the introduction say why you think the project is worth doing and show how it relates to what is already known about the topic. State the immediate and longer-term aims of the work. Then describe the experimental design and the procedures and protocols you will use. Mention ethical rules or codes of practice that will be followed. Don't describe the methods in detail unless the funding body's guidelines require you to do so. In appropriate fields, outline the results you expect to get and discuss their potential significance. Include tables and figures where necessary, provided that their number or the overall length of the proposal stays within the stipulated limits.

Don't make exaggerated claims when you describe your research plan.

Say whether the project could be included in or is relevant to international projects such as the International Biological Programme (IBP) or the International Geological Correlation Programme (IGCP).

Don't make exaggerated claims when you describe your research plan. Assessors are not impressed by an applicant who seems to have his or her head in the clouds. Do point to different paths the project could take if the results go against your expectations or if they open up additional interesting possibilities. The proposal should be both flexible and specific.

Drafting the budget

The budget should be 'reasonable, believable, well-researched, and superbly justified' (Reif-Lehrer 1982). Preparing it calls for just as much care as preparing the title, abstract, and research plan. You will usually be asked for a detailed list of proposed expenditure on personnel, equipment, supplies, travel, and other items, with or without a summary version of the costs. If you are applying for a small grant, a summary version of the budget may be sufficient. Don't forget to allow for any tax problems you or others working on the project might run into. If field work is involved, point this out; funding bodies may have different allowances for different places.

For a detailed budget, cost each item carefully. Allow for inflation or exchange-rate fluctuations and say how you do so. Write down all the totals (don't expect the assessors to do the arithmetic for you), and make sure you do the sums correctly (do expect at least one assessor to be quick on the draw with a calculator). A spreadsheet program will be useful if you are drafting the budget on a computer.

You must also explain the necessity for each category or each major item of expenditure. If you are asking for extra equipment, say what equipment is already available in the department and why it is insufficient for the project. If you want money for travel to meetings or to other laboratories, explain why such travel is necessary to the success of the project. Where appropriate, quote corroborating statements from departmental heads or supervisors and from the heads of any institutes you plan to visit.

If you are asking for support for research associates or assistants, name them if the funding body requires you to do so. Explain what their part in the investigation will be and why their assistance is essential. Justify their rank – for example, why is a postdoctoral assistant necessary rather than a postgraduate?

Assembling a coherent final proposal

Put the draft of your proposal away for a while before rereading and rewriting it. Persuade suitable colleagues to criticize the proposal, including the budget, before you prepare the final version. Make sure that spelling, grammar, reference details, and arithmetic are correct. Prepare curricula vitae in the way described below.

Reread the guidelines and application forms before the final proposal is typed. Fill in forms carefully. Make sure that forms and any additional pages are typed clearly and boldly enough to produce good photocopies. Check and correct the typing before you make the necessary number of copies (including a copy for yourself).

Separate the photocopies into sets, add any other items such as reprints that are required, and check that each set is complete. Put an elastic band round each set, or put each set in a folder. Finally, write a brief covering letter and put the letter and the necessary number of sets of the proposal into a strong envelope or box for mailing to the funding body. If you want the safe arrival of your application to be acknowledged you may need to enclose a stamped addressed envelope or an acknowledgement card (sometimes supplied by the funding body). If the application has to arrive by a certain date, make sure you post it in good time – use express mail or a courier service if necessary. If you are sending a package to another country, beware of delays in customs, and include a customs declaration, if necessary.

PREPARING A CURRICULUM VITAE FOR
A GRANT PROPOSAL

Prepare your curriculum vitae (CV) for a grant proposal as carefully as you prepare the rest of the proposal. For many proposals the information required in CVs will be laid down in the application forms. The same information may be required for any research associates or research assistants included in the grant proposal.

The format of the CV is also often dictated by the application form. If not, there are various ways you could prepare it. Follow the pattern most acceptable to the funding body, if you know what that pattern is; otherwise, observe national custom in the funding body's country. If you have no model to follow, the order suggested here is fairly standard.

Start with your name, institutional address, and date of birth. Name your university qualifications and say where, when, and in what subject you obtained them; include relevant information such as the title of your doctoral thesis. List the posts you have held, with dates. Include guest lectureships, international prizes, or invited membership of national or international committees or organizations. Mention participation in international meetings relevant to the topic of your project.

List your publications in the CV if you are not asked to include them elsewhere in the application. Include publications in preparation and say what stage they have reached. If your name appears on a lot of publications, either mark the most important ones with an asterisk (remember to explain what the asterisk means) or leave out the less important ones but refer to their existence. Produce a complete list as an appendix if the number is large enough to justify this, or simply state the number.

Arrange all the information in a way that is easy to read (see Fig. 15.1). The CV should preferably fit on one or two A4 or $8\frac{1}{2} \times 11$-inch pages – but this will depend on how many jobs you have held and how many articles you have published. Type the CV carefully, on one side of the paper only, preferably using a word processor. If you have it typed professionally, note that some firms specializing in CV preparation produce glossy but insubstantial CVs that are unsuitable for grant proposals and academic jobs. Whoever types the CV, check it carefully before sending it to a funding body.

PREPARING A CURRICULUM VITAE FOR A
JOB APPLICATION

A CV accompanying a job application (Fig. 15.1) usually includes more details than one forming part of a grant proposal. Add your home address to the information described above. If required, or if customary in your discipline or country, give details of your education before you started university (or equivalent) education. Include any awards or prizes received during your education or career to date. Name any important extracurricular activities relevant to the job you are applying for. If you took time off to walk round the world or for some other purpose, explain the gap in your career. If you don't do this, some assessors will suspect criminality or insanity.

· When appropriate, describe your responsibilities in your present or most recent job. There is no need to state your present salary unless you want to or are asked to do so.

At the end of the CV give the names, addresses, and telephone and fax numbers of the required number of referees – but first obtain their written permission to name them for this purpose and make sure they remember when you studied with them or worked for them. Choose referees of good reputation and standing; they should of course be people who are likely to give you a good reference. Tell them what job you are applying for and send them a copy of your application or at least of your CV to help them when they write the reference. If you cite your present head of department as a referee you may want to add 'Do not contact without my permission'. Don't send written references unless you are asked to do so, and then send copies, not the originals.

Send a brief covering letter with the CV, naming the job you are applying for and saying where you saw it advertised. Type the letter unless a handwritten one is requested. Consider whether to enclose a photograph if one is requested. If necessary, ask the organization to which you are applying not to contact your present employer without first contacting you. Don't enclose copies of your publications unless you are asked for these.

Type the CV carefully, as described above, or have it prepared by someone who specializes in producing CVs. If possible, keep a copy on disk so that you can update it easily.

Good luck with the project, or the job!

Curriculum vitae

Janet T Brown, PhD
Department of Geophysics
University of the Lowlands
Downtown, Downshire 6XY Z9A
Tel. 0123-456789
Fax 0123-457234

Age: 30
Date of birth: 1 May 1960
Health: excellent
Home address:
99 Prince Street
Downtown, Downshire 3XY Z2B
Tel. 0123-987654

Education

1971–1977: Uptown High School, Uptown
1977–1981: Midlands University, Chesterham
1981: BSc, 1st-class honours in geology
1981–1984: Dept of Geophysics, University of Camford
1984: PhD conferred
Thesis title: Melting in silicate rocks in the Lower Upshire region
Gatenew Prize awarded

Positions held

1984–1987: Associate Lecturer in Geophysics, Brighton University, Brighton, NY,
USA
1987–present: Lecturer in Geophysics, University of the Lowlands, Downtown,
Downshire, UK

Publications

Brown JT. 1984. Plastic deformation of quartz in deep seismic sounding sections
in Upper New York State. Brighton Journal of Geophysical Research 23:22–28.
Brown JT, Smith D, Jones S. 1985. Dependence of flow temperature on differen-
tial stress in quartz and olivine. Journal of the New York Academy of Geophysics
120:1055–1057.

[etc.]

Other interests

Member of the Downtown Deep Explorers Club (Hon. Secretary since 1989)

Referees

Professor M Hayman
Dept of Geology
Midlands University
Chesterham, C1D A3B, UK
Tel. 0123-456781
Fax 0123-457234

Dr S Laysmith
Institute of Geophysics
Brighton University
Brighton, NY 99999-1111, USA
Tel. 222-333 4444;
Fax 222-333 5555

Figure 15.1 A curriculum vitae to accompany a job application. More detailed
background information such as a description of what the current job involves
should be added in appropriate places. Lay-out can differ from that shown here
provided that the information is in a logical order and attractively presented.

SUMMARY

(1) Formulate a question worth answering;   (2) Decide which funding body to apply to;   (3) Obtain the funding body's instructions and application forms;   (4) Start preparing your application in good time;   (5) Draft the title, abstract, and research plan; (6) Draft the budget and explain why each major item is needed; (7) Prepare the final proposal and type and check it carefully; (8) Prepare a curriculum vitae to go with a proposal;   (9) Prepare a curriculum vitae for a job application.

APPENDIX ONE

*Uniform requirements for manuscripts submitted to biomedical journals**

INTERNATIONAL COMMITTEE OF MEDICAL
JOURNAL EDITORS

In January 1978 a group of editors from some major biomedical journals published in English met in Vancouver, British Columbia, and decided on uniform technical requirements for manuscripts to be submitted to their journals. These requirements, including formats for bibliographic references developed for the Vancouver group by the National Library of Medicine, were published in three of the journals early in 1979. The Vancouver group evolved into the International Committee of Medical Journal Editors. At the October 1981 meeting the requirements were revised slightly and published in a second edition in 1982. Since then the group has issued several separate statements, and these have been incorporated into the main part of this, the third, edition.

Over 300 journals have agreed to receive manuscripts prepared in accordance with the initial, previously published, requirements. It is important to emphasise what these requirements imply and what they do not.

Firstly, the requirements are instructions to authors on how to prepare manuscripts, not to editors on publication style. (But many journals have drawn on these requirements for elements of their publication styles.)

Secondly, if authors prepare their manuscripts in the style specified in these requirements editors of the participating journals will not return manuscripts for changes in these details of style. Even so, manuscripts may be altered by journals to conform with details of their own publication styles.

Thirdly, authors sending manuscripts to a participating journal should

*Reproduced from British Medical Journal 1988;296:401–405.

not try to prepare them in accordance with the publication style of that journal but should follow the "Uniform requirements for manuscripts submitted to biomedical journals."

Nevertheless authors must also follow the instructions to authors in the journal as to what topics are suitable for that journal and the types of papers that may be submitted – for example, original articles, reviews, case reports. In addition, the journal's instructions are likely to contain other requirements unique to that journal, such as number of copies of manuscripts, acceptable languages, length of articles, and approved abbreviations.

Participating journals are expected to state in their instructions to authors that their requirements are in accordance with the "Uniform requirements for manuscripts submitted to biomedical journals" and to cite a published version.

This document will be revised at intervals. Inquiries and comments from Central and North America about these requirements should be sent to Edward J Huth, MD, *Annals of Internal Medicine* [American College of Physicians, Independence Mall West, Sixth Street at Race, Philadelphia, PA 19106-1572, USA]; those from other regions should be sent to Stephen Lock, MD, *British Medical Journal*, British Medical Association, Tavistock Square, London WC1H 9JR, United Kingdom. Note that these two journals provide secretariat services for the International Committee of Medical Journal Editors; they do not handle manuscripts intended for other journals. Papers intended for other journals should be sent directly to the offices of those journals.

SUMMARY OF REQUIREMENTS

Type the manuscript double spaced, including title page, abstract, text, acknowledgments, references, tables, and legends.

Each manuscript component should begin on a new page, in the following sequence: title page; abstract and key words; text; acknowledgments; references; tables (each table complete with title and footnotes on a separate page); and legends for illustrations.

Illustrations must be good quality, unmounted glossy prints, usually 127×173 mm (15×7 in) but no larger than 203×254 mm (8×10 in).

Submit the required number of copies of manuscript and figures (see journal's instructions) in a heavy paper envelope. The submitted manuscript should be accompanied by a covering letter, as described under "Submission of manuscripts," and permissions to reproduce previously published material or to use illustrations that may identify human subjects.

Follow the journal's instructions for transfer of copyright. Authors should keep copies of everything submitted.

PRIOR AND DUPLICATE PUBLICATION

Most journals do not wish to consider for publication a paper on work that has already been reported in a published paper or is described in a paper submitted or accepted for publication elsewhere. This policy does not usually preclude consideration of a paper that has been rejected by another journal or of a complete report that follows publication of a preliminary report, usually in the form of an abstract. When submitting a paper an author should always make a full statement to the editor about all submissions and previous reports that might be regarded as prior or duplicate publication of the same or very similar work. Copies of such material should be included with the submitted paper to help the editor decide how to deal with the matter.

Multiple publication – that is, the publication more than once of the same study, irrespective of whether the wording is the same – is rarely justified. Secondary publication in another language is one possible justification, provided the following conditions are met.

(1) The editors of both journals concerned are fully informed; the editor concerned with secondary publication should have a photocopy, reprint, or manuscript of the primary version.
(2) The priority of the primary publication is respected by a publication interval of at least two weeks.
(3) The paper for secondary publication is written for a different group of readers and is not simply a translated version of the primary paper; an abbreviated version will often be sufficient.
(4) The secondary version reflects faithfully the data and interpretations of the primary version.
(5) A footnote on the title page of the secondary version informs readers, peers, and documenting agencies that the paper was edited, and is being published, for a national audience in parallel with a primary version based on the same data and interpretations. A suitable footnote might read as follows: "This article is based on a study first reported in the [title of journal, with full reference]."

Multiple publication other than as defined above is not acceptable to editors. If authors violate this rule they may expect appropriate editorial action to be taken.

Preliminary release, usually to public media, of scientific information described in a paper that has been accepted but not yet published is a violation of the policies of many journals. In a few cases, and only by arrangement with the editor, preliminary release of data may be acceptable – for example, to warn the public of health hazards.

PREPARATION OF MANUSCRIPT

Type the manuscript on white bond paper, 216×279 mm (8½×11 in) or ISO A4 (212×297 mm), with margins of at least 25 mm (1 in). Type only on one side of the paper. Use double spacing throughout, including title page, abstract, text, acknowledgments, references, tables, and legends for illustrations. Begin each of the following sections on separate pages: title page, abstract and key words, text, acknowledgments, references, individual tables, and legends. Number pages consecutively, beginning with the title page. Type the page number in the upper or lower righthand corner of each page.

Title page

The title page should carry (a) the title of the article, which should be concise but informative; (b) first name, middle initial, and last name of each author, with highest academic degree(s) and institutional affiliation; (c) name of department(s) and institution(s) to which the work should be attributed; (d) disclaimers, if any; (e) name and address of author responsible for correspondence about the manuscript; (f) name and address of author to whom requests for reprints should be addressed or statement that reprints will not be available from the author; (g) source(s) of support in the form of grants, equipment, drugs, or all of these; and (h) a short running head or footline of no more than 40 characters (count letters and spaces) placed at the foot of the title page and identified.

Authorship

All persons designated as authors should qualify for authorship. Each author should have participated sufficiently in the work to take public responsibility for the content.

Authorship credit should be based only on substantial contributions to (a) conception and design, or analysis and interpretation of data; and to (b) drafting the article or revising it critically for important intellectual content; and on (c) final approval of the version to be published. Conditions (a), (b), and (c) must all be met. Participation solely in the acquisition of funding or the collection of data does not justify authorship. General supervision of the research group is also not sufficient for authorship. Any part of an article critical to its main conclusions must be the responsibility of at least one author.

A paper with corporate (collective) authorship must specify the key persons responsible for the article; others contributing to the work should be recognised separately (see "Acknowledgments").

Editors may require authors to justify the assignment of authorship.

Abstract and key words

The second page should carry an abstract of no more than 150 words. The abstract should state the purposes of the study or investigation, basic procedures (selection of study subjects or experimental animals; observational and analytical methods), main findings (give specific data and their statistical significance, if possible), and the principal conclusions. Emphasise new and important aspects of the study or observations.

Below the abstract provide, and identify as such, three to 10 key words or short phrases that will assist indexers in cross indexing the article and may be published with the abstract. Use terms from the medical subject headings (MeSH) list of *Index Medicus*; if suitable MeSH terms are not yet available for recently introduced terms present terms may be used.

Text

The text of observational and experimental articles is usually – but not necessarily – divided into sections with the headings introduction, methods, results, and discussion. Long articles may need subheadings within some sections to clarify their content, especially the results and discussion sections. Other types of articles such as case reports, reviews, and editorials are likely to need other formats. Authors should consult individual journals for further guidance.

Introduction – State the purpose of the article. Summarise the rationale for the study or observation. Give only strictly pertinent references, and do not review the subject extensively. Do not include data or conclusions from the work being reported.

Methods – Describe your selection of the observational or experimental subjects (patients or experimental animals, including controls) clearly. Identify the methods, apparatus (manufacturer's name and address in parentheses), and procedures in sufficient detail to allow other workers to reproduce the results. Give references to established methods, including statistical methods (see below); provide references and brief descriptions for methods that have been published but are not well known; describe new or substantially modified methods, give reasons for using them, and evaluate their limitations. Identify precisely all drugs and chemicals used, including generic name(s), dose(s), and route(s) of administration.

Ethics – When reporting experiments on human subjects indicate whether the procedures followed were in accordance with the ethical standards of the responsible committee on human experimentation (institutional or regional) or with the Helsinki Declaration of 1975, as revised in 1983. Do not use patients' names, initials, or hospital numbers, especially in any illustrative material. When reporting experiments on animals indicate whether the institution's or the National Research Council's guide for, or any national law on, the care and use of laboratory animals was followed.

Statistics – Describe statistical methods with enough detail to enable a knowledgeable reader with access to the original data to verify the reported results. When possible quantify findings and present them with appropriate indicators of measurement error or uncertainty (such as confidence intervals). Avoid sole reliance on statistical hypothesis testing, such as the use of p values, which fails to convey important quantitative information. Discuss eligibility of experimental subjects. Give details about randomisation. Describe the methods for, and success of, any blinding of observations. Report treatment complications. Give numbers of observations. Report losses to observation (such as dropouts from a clinical trial). References for study design and statistical methods should be to standard works (with pages stated) when possible rather than to papers where designs or methods were originally reported. Specify any general use computer programs used.

Put general descriptions of methods in the methods section. When data are summarised in the results section specify the statistical methods used to analyse them. Restrict tables and figures to those needed to explain the argument of the paper and to assess its support. Use graphs as an alternative to tables with many entries; do not duplicate data in graphs and tables. Avoid non-technical uses of technical terms in statistics, such as "random" (which implies a randomising device), "normal," "significant," "correlations," and "sample." Define statistical terms, abbreviations, and most symbols.

Results – Present your results in logical sequence in the text, tables, and illustrations. Do not repeat in the text all the data in the tables or illustrations, or both; emphasise or summarise only important observations.

Discussion – Emphasise the new and important aspects of the study and the conclusions that follow from them. Do not repeat in detail data or other material given in the introduction or the results section. Include in the discussion section the implications of the findings and their limitations, including implications for future research. Relate the observations to other relevant studies. Link the conclusions with the goals of the study but avoid unqualified statements and conclusions not completely supported by your data. Avoid claiming priority and alluding to work that has not been completed. State new hypotheses when warranted, but clearly label them as such. Recommendations, when appropriate, may be included.

Acknowledgments

At an appropriate place in the article (title page footnote or appendix to the text; see the journal's requirement) one or more statements should specify (a) contributions that need acknowledging but do not justify authorship, such as general support by a departmental chairman; (b) acknowledgments of technical help; (c) acknowledgments of financial and material support, specifying the nature of the support; (d) financial relationships that may pose a conflict of interest.

Persons who have contributed intellectually to the paper but whose contributions do not justify authorship may be named and their function or contribution described – for example, "scientific adviser," "critical review of study proposal," "data collection," "participation in clinical trial." Such persons must have given their permission to be named. Authors are responsible for obtaining written permission from persons acknowledged by name because readers may infer their endorsement of the data and conclusions.

Technical help should be acknowledged in a paragraph separate from those acknowledging other contributions.

References

Number references consecutively in the order in which they are first mentioned in the text. Identify references in text, tables, and legends by arabic numerals in parentheses. References cited only in tables or in legends should be numbered in accordance with a sequence established by the first identification in the text of the particular table or illustration.

Use the style of the examples below, which are based on the formats used by the US National Library of Medicine in *Index Medicus*. The titles of journals should be abbreviated according to the style used in *Index Medicus*. Consult *List of Journals Indexed in Index Medicus*, published annually as a separate publication by the library and as a list in the January issue of *Index Medicus*; also see the list of journal titles and abbreviated titles at the end of this document [not included in this book].

Try to avoid using abstracts as references; "unpublished observations" and "personal communications" may not be used as references, although references to written, not oral, communications may be inserted (in parentheses) in the text. Include among the references papers accepted but not yet published; designate the journal and add "in press" (in parentheses). Information from manuscripts submitted but not yet accepted should be cited in the text as "unpublished observations" (in parentheses).

The references must be verified by the author(s) against the original documents.

Examples of correct forms of references are given below.

JOURNALS

(1) *Standard journal article* – (List all authors when six or less; when seven or more, list only first three and add et al.) You CH, Lee KY, Chey RY, Menguy R. Electrogastrographic study of patients with unexplained nausea, bloating and vomiting. Gastroenterology 1980;79:311–4.

(2) *Corporate author* The Royal Marsden Hospital Bone-Marrow Transplantation Team. Failure of syngeneic bone-marrow graft without preconditioning in post-hepatitis marrow aplasia. Lancet 1977;ii:242–4.

(3) *No author given* Anonymous. Coffee drinking and cancer of the pancreas [Editorial]. Br Med J 1981;283:628.

(4) *Journal supplement* Mastri AR. Neuropathy of diabetic neurogenic bladder. Ann Intern Med 1980;92(2 Pt 2):316–8.
Frumin AM, Nussbaum J, Esposito M. Functional asplenia: demonstration of splenic activity by bone marrow scan [Abstract]. Blood 1979;54 (suppl 1):26a.

(5) *Journal paginated by issue* Seaman WB. The case of the pancreatic pseudocyst. Hosp Pract 1981;16(Sep):24–5.

BOOKS AND OTHER MONOGRAPHS

(6) *Personal author(s)* Eisen HN. Immunology: an introduction to molecular and cellular principles of the immune response. 5th ed. New York: Harper and Row, 1974:406.

(7) *Editor, compiler, chairman as author* Dausset J, Colombani J, eds. Histocompatibility testing 1972. Copenhagen: Munksgaard, 1973:12–8.

(8) *Chapter in a book* Weinstein L, Swartz MN. Pathogenic properties of invading micro-organisms. In: Sodeman WA Jr, Sodeman WA, eds. Pathologic physiology: mechanisms of disease. Philadelphia: W B Saunders, 1974:457–72.

(9) *Published proceedings paper* DuPont B. Bone marrow transplantation in severe combined immunodeficiency with an unrelated MLC compatible donor. In: White HJ, Smith R, eds. Proceedings of the third annual meeting of the International Society for Experimental Hematology. Houston: International Society for Experimental Hematology, 1974:44–6.

(10) *Monograph in a series* Hunninghake GW, Gadek JE, Szapiel SV, et al. The human alveolar macrophage. In: Harris CC, ed. Cultured human cells and tissues in biomedical research. New York: Academic Press, 1980:54–6. (Stoner GD, ed. Methods and perspectives in cell biology; vol 1.)

(11) *Agency publication* Ranofsky AL. Surgical operations in short-stay hospitals: United States – 1975. Hyattsville, Maryland: National Center for Health Statistics, 1978; DHEW publication no. (PHS)78-1785. (Vital and health statistics; series 13; no. 34.)

(12) *Dissertation or thesis* Cairns RB. Infrared spectroscopic studies of solid oxygen [Dissertation]. Berkeley, California: University of California, 1965. 156 pp.

OTHER ARTICLES

(13) *Newspaper article* Shaffer RA. Advances in chemistry are starting to unlock mysteries of the brain: discoveries could help cure alcoholism and insomnia, explain mental illness. How the messengers work. Wall Street Journal 1977 Aug 12:1(col 1),10(col 1).

(14) *Magazine article* Roueche B. Annals of medicine: the Santa Claus culture. The New Yorker 1971 Sep 4:66–81.

Tables

Type each table double spaced on a separate sheet. Do not submit tables as photographs. Number tables consecutively in the order of their first citation in the text and supply a brief title for each. Give each column a short or abbreviated heading. Place explanatory matter in footnotes, not in the heading. Explain in footnotes all non-standard abbreviations that are used in each table. For footnotes use the following symbols, in this sequence: *, †, ‡, §, ||, ¶, **, ††, . . .

Identify statistical measures of variations such as standard deviation and standard error of the mean.

Do not use internal horizontal and vertical rules.

Be sure that each table is cited in the text.

If you use data from another published or unpublished source obtain permission and acknowledge fully.

The use of too many tables in relation to the length of the text may produce difficulties in the layout of pages. Examine issues of the journal to which you plan to submit your paper to estimate how many tables can be used per 1000 words of text.

The editor, on accepting a paper, may recommend that additional tables containing important back up data too extensive to publish be deposited with an archival service, such as the National Auxiliary Publication Service in the United States, or made available by the authors. In that event an appropriate statement will be added to the text. Submit such tables for consideration with the paper.

Illustrations

Submit the required number of complete sets of figures. Figures should be professionally drawn and photographed; freehand or typewritten lettering is unacceptable. Instead of original drawings, roentgenograms, and other material send sharp, glossy black and white photographic prints, usually 127×173 mm (5×7 in) but no larger than 203×254 mm (8×10 in). Letters, numbers, and symbols should be clear and even throughout and of sufficient size that when reduced for publication each item will still be legible. Titles and detailed explanations belong in the legends for illustrations, not on the illustrations themselves.

Each figure should have a label pasted on its back indicating the number of the figure, author's name, and top of the figure. Do not write on the back of figures or scratch or mar them by using paper clips. Do not bend figures or mount them on cardboard.

Photomicrographs must have internal scale markers. Symbols, arrows, or letters used in the photomicrographs should contrast with the background.

If photographs of persons are used either the subjects must not be identifiable or their pictures must be accompanied by written permission to use the photograph.

Figures should be numbered consecutively according to the order in which they have been first cited in the text. If a figure has been published acknowledge the original source and submit written permission from the copyright holder to reproduce the material. Permission is required irrespective of authorship or publisher, except for documents in the public domain.

For illustrations in colour, ascertain whether the journal requires colour negatives, positive transparencies, or colour prints. Accompanying drawings marked to indicate the region to be reproduced may be useful to the editor. Some journals publish illustrations in colour only if the author pays for the extra cost.

Legends for illustrations

Type legends for illustrations double spaced, starting on a separate page, with arabic numerals corresponding to the illustrations. When symbols, arrows, numbers, or letters are used to identify parts of the illustrations identify and explain each one clearly in the legend. Explain the internal scale and identify method of staining in photomicrographs.

UNITS OF MEASUREMENT

Measurements of length, height, weight, and volume should be reported in metric units (metre, kilogram, litre) or their decimal multiples.

Temperatures should be given in degrees Celsius. Blood pressures should be given in milligrams of mercury.

All haematological and clinical chemistry measurements should be reported in the metric system in terms of the International System of Units (SI). Editors may request that alternative or non-SI units be added by the authors before publication.

ABBREVIATIONS AND SYMBOLS

Use only standard abbreviations. Avoid abbreviations in the title and abstract. The full term for which an abbreviation stands should precede its first use in the text unless it is a standard unit of measurement.

SUBMISSION OF MANUSCRIPTS

Mail the required number of manuscript copies in a heavy paper envelope, enclosing the manuscript copies and figures in cardboard, if necessary, to prevent bending of photographs during mail handling. Place photographs and transparencies in a separate heavy paper envelope.

Manuscripts must be accompanied by a covering letter. This must include (a) information on prior or duplicate publication or submission elsewhere of any part of the work; (b) a statement of financial or other relationships that might lead to a conflict of interests; (c) a statement that the manuscript has been read and approved by all authors; and (d) the name, address, and telephone number of the corresponding author, who is responsible for communicating with the other authors about revisions and final approval of the proofs. The letter should give any additional information that may be helpful to the editor, such as the type of article in the particular journal the manuscript represents and whether the author(s) will be willing to meet the cost of reproducing colour illustrations.

The manuscript must be accompanied by copies of any permissions to reproduce published material, to use illustrations or report sensitive personal information of identifiable persons, or to name persons for their contributions.

. . .

Citations of this document should be to one of the sources listed below:

International Committee of Medical Journal Editors. Uniform requirements for manuscripts submitted to biomedical journals. Ann Intern Med 1988;108:258–65.

International Committee of Medical Journal Editors. Uniform requirements for manuscripts submitted to biomedical journals. Br Med J 1988;296:401–5.

This document is not covered by copyright: it may be copied or reprinted without permission.

APPENDIX TWO

*Terms to avoid**

Note that the terms in the left-hand column are not wrong, merely longer or more pompous than the suggested alternatives. Choose terms from the right-hand column in preference to those from the left, but use those from the left-hand column when necessary ('approximately' and 'a majority', for example, may be more accurate in a particular context than the alternatives shown here).

Long or (sometimes) wrong	Better choice (often)
a majority of	most
a number of	few, several, many
accounted for by the fact that	because
and moreover	moreover
an order of magnitude	ten times
anticipate	expect
approximately	about
are of the same opinion	agree
as a consequence of	because
as already stated	[omit]
as can be seen from Figure 1, substance Z reduces twitching	substance Z reduces twitching (Fig. 1)
as far as these experiments are concerned, they show	these experiments show
as of now	now, from now on
as regards this species, it	this species is
as to whether	whether
as yet	yet
at a later date	later
at some future time	later
attempt	try
at the end of the day	[omit]

*Including entries borrowed from O'Connor & Woodford (1975), Huth (1987), and CBE Style Manual Committee (1983).

Long or (sometimes) wrong	Better choice (often)
at the present moment, at this point in time	now
bright yellow in colour	bright yellow
by means of	by, with
case	patient
caused damage to	damaged
commence	begin, start
completely filled	filled
conducted inoculation experiments on	inoculated
consensus of opinion	consensus
considerable amount of	much
considerable number of	many, most
decreased number of	fewer, less
decreased relative to	less than, lower than
definitely proved	proved
despite the fact that	although
due to the fact that	because
during the course of	during, while
during the time that	while
elevated	raised, higher, more
employ	use
encountered	met
equivalent as far as acceptability is concerned	equally acceptable
fewer in number	fewer
following (e.g. an event)	after
for the reason that	because
from the standpoint of	according to
fully cognizant of the fact that	aware that
goes under the name of	is called
has the capability of	can, is able to
having regard to	about
if conditions are such that	if
in a considerable number of cases	often
in all cases	always, invariably
in close proximity to	close to
in connection with	about, concerning
in excess of	more than, above
in order to	to

Long or (sometimes) wrong	_Better choice (often)_
in regard to, in relation to, in respect of, in the case of, etc.	use in, for, about, or with, or omit, as appropriate
in the event that	if
in the present communication	here; in this paper
in view of the fact that	because
integral part	part
it is of interest to note that	[omit]
it may, however, be noted that	but
join together	join
large numbers of	many
lazy in character	lazy
major breakthrough	breakthrough
mass media	media
methodology	methods
multiple	several, different
my personal opinion	my opinion
on the basis of	because, by, from
owing to the fact that	because
oval in shape	oval
paradigm	example, pattern
parameter	index, criterion, measure, value
permeate throughout	permeate
penetrate into	penetrate
pertaining to	on, about
plethora	too many
prior to	before
reported to the effect that	reported that
similar in every detail	the same
a serious malfunction has occurred in the system	the system has failed
subsequent to	after
take into consideration	consider
temporary reprieve	reprieve
terminate	end
the test in question	this test
the tests have not as yet	the tests have not
the treatment having been performed	after treatment
therapeutic treatment	treatment
there can be little doubt that this is	this is probably

Long or (sometimes) wrong	*Better choice (often)*
there is, there are	[often unnecessary; reword the sentence]
there is a lot of care that goes into	much care goes into
thorough investigation	investigation
throughout the whole of the book	throughout the book
to an extent equal to that of X	as much as X
two equal halves	halves
upon	on
utilize	use
very, quite, rather, and other vague qualifiers	[omit]
vague words, such as area, character, conditions, field, level, nature, problem, process, situation, structure, system	change to more precise words appropriate to the context
when and if	if
whether or not	whether
with reference to, with regard to	about [or omit]

References

Altman D G, Gore S M, Gardner M J, Pocock S J. 1983. Statistical guidelines for contributors to medical journals. British Medical Journal 286:1489–1493.

American Institute of Physics. 1978. Style manual. New York: American Institute of Physics.

American Psychological Association. 1983. Publication manual of the American Psychological Association, 3rd ed. Washington, DC: American Psychological Association.

American Society of Agronomy et al. 1984. Handbook and style manual. Madison, WI: American Society of Agronomy, Crop Science Society of America, Soil Science Society of America.

American Society for Microbiology. 1985. ASM style manual for journals and books. Washington, DC: American Society for Microbiology.

ANSI Z39.22:1981. American national standard for proof corrections. New York: American National Standards Institute.

Bates R L, Jackson J A. 1987. Glossary of geology, 3rd ed. Alexandria, VA: American Geological Institute.

Benjamin B. 1989. Elementary primer of English grammar. London: Futura.

Bishop A. 1984. Slides – planning and producing slide programs. Rochester, NY: Eastman Kodak (Publication S-30). 159 pp. [Cited by Heron 1989.]

Booth V. 1985. Communicating in science: writing and speaking. Cambridge: Cambridge University Press.

Briscoe M H. 1990. Researchers' guide to scientific and medical illustration. New York: Springer.

Brouwer O. 1983. The cartographer's role and requirements. Scholarly Publishing 14:231–242.

Bryant M, Cox S. 1983. The editor and the illustration. Scholarly Publishing 14:213–230.

BS 1629:1989. Recommendations for references to published materials. London: British Standards Institution.

BS 4148:1970–1975. Abbreviation of titles of periodicals. Part 1: Principles, 1970; Part 2: Word-abbreviation list, 1975. London: British Standards Institution.

BS 5261:1976. Guide to copy preparation and proof corrections. Part 2: Specifications for typographic requirements, marks for copy preparation and proof correction, proofing procedure. London: British Standards Institution.

Butcher J. 1983. Copy-editing: the Cambridge handbook, 2nd ed. Cambridge: Cambridge University Press.

Carey G V. 1971. Mind the stop: brief guide to punctuation. Harmondsworth, UK: Penguin Books.

CBE Scientific Illustration Committee. 1988. Illustrating science: standards for publication. Bethesda, MD: Council of Biology Editors (9650 Rockville Pike, Bethesda, MD 20814-3929).

CBE Style Manual Committee. 1983. CBE style manual: a guide for authors, editors, and publishers in the biological sciences, 5th ed. Bethesda, MD: Council of Biology Editors.

Cleveland W S. 1985. The elements of graphing data. Monterey, CA: Wadsworth.

Cochran W, Fenner P, Hill M (eds). 1984. Geowriting: a guide to writing, editing, and printing in earth science, 4th ed. Alexandria, VA: American Geological Institute.

Day R A. 1988. How to write and publish a scientific paper, 3rd ed. Phoenix, AZ: Oryx Press & Cambridge: Cambridge University Press.

DeBakey L. 1976. The scientific journal: editorial policies and practices – guidelines for editors, reviewers, and authors. St Louis, MO: Mosby.

DIN 16 511. 1966. Korrekturzeichen. Berlin: Deutsches Institut für Normung.

Dodd J S, ed. 1986. The ACS style guide: a manual for authors and editors. Washington, DC: American Chemical Society.

Dutro J T, Jr, Dietrich R V, Foose R M, compilers. 1989. AGI data sheets for geology in the field, laboratory, and office. Alexandria, VA: American Geological Institute.

Eastman Kodak. 1977. Basic color for the graphic arts. Rochester, NY: Eastman Kodak Co. (Publication Q-7.) [Cited by CBE Scientific Illustration Committee 1988. Other Eastman Kodak publications may also be useful.]

Ebel H F, Bliefert C, Russey W E. 1987. The art of scientific writing. Weinheim, FRG: VCH.

ELSE [European Life Science Editors]–Ciba Foundation Workshop. 1978. References in scientific publications. Earth & Life Science Editing No. 7:18–21. [Also in O'Connor 1978, p. 176–185.]

Evans M. 1978. The abuse of slides. British Medical Journal 1:905–908. [Also in: How to do it, 2nd ed. London: British Medical Association, p. 152–160.]

Follett W. 1974. Modern American usage. New York: Warner.

Fowler H W, revised by Sir Ernest Gowers. 1965. A dictionary of modern English usage, 2nd ed. Oxford: Oxford University Press.

Gardner M J, Altman D G, eds. 1989. Statistics with confidence. London: British Medical Association.

Gordon K. 1984. The transitive vampire: a handbook of grammar for the innocent, the eager, and the doomed. New York: Times Books & London: Severn House.

Gowers E, revised by S. Greenbaum & J. Whitcut. 1986. The complete plain words, 3rd ed. London: Her Majesty's Stationery Office & Harmondsworth, Middlesex: Penguin Books.

Haynes R B, McKibbon K A, Fitzgerald D, Guyatt G G, Walker C J, Sackett D L. 1986. How to keep up with the medical literature: I. Why try to keep up and how to get started. Annals of Internal Medicine 105:149–153.

Heron D. 1989. Preparing and presenting a slide talk. AGI data sheet 88.1. In: Dutro J T, Jr, Dietrich R V, Foose R M, compilers. AGI data sheets for geology in the field, laboratory, and office. Alexandria, VA: American Geological Institute.

215

Howard P. 1984. The state of the language. London: Hamish Hamilton.

Huguier H, Maisonneuve H, Benhamou C L et al. 1990. La rédaction médicale. Paris: Doin.

Hull R, Brown F, Payne C. 1989. Directory & dictionary of animal, bacterial and plant viruses. London: Macmillan.

Huth E J. 1987. Medical style and format: an international manual for authors, editors, and publishers. Philadelphia: ISI Press [now available from Williams & Wilkins, Baltimore, MD].

ICMJE [International Committee of Medical Journal Editors]. 1988. Uniform requirements for manuscripts submitted to biomedical journals. British Medical Journal 296:401–405/Annals of Internal Medicine 108:258–265. [See Appendix 1.]

Institute of Physics. 1983. Notes for authors. London: Institute of Physics.

International Anatomical Nomenclature Committee/Subcommittees. 1989. Nomina Anatomica, 6th ed; Nomina Histologica, 3rd ed; Nomina Embryologica, 3rd ed. Edinburgh: Churchill Livingstone.

ISDS [International Centre for the Registration of Serials]. 1975. International list of periodical title word abbreviations [formerly ISO 833]. Paris: ISDS [20 rue Bachaumont, 75002 Paris, France].

ISO 4:1984. Documentation – rules for the abbreviation of title words and titles of publications. Geneva: International Organization for Standardization. [ISO, Case postale 56, CH-1211 Geneva 20, Switzerland.]

ISO 31-11:1978. Mathematical signs and symbols for use in the physical sciences and technology. Geneva: ISO.

ISO 214:1976. Documentation – abstracts for publications and documentation. Geneva: ISO.

ISO 215:1986. Documentation – presentation of contributions to periodicals and other serials. Geneva: ISO.

ISO 690:1987. Documentation – bibliographic references – content, form and structure. Geneva: ISO.

ISO 2384:1977. Documentation – presentation of translations. Geneva: ISO.

ISO 5807:1985. Information processing – documentation symbols and conventions for data, program and system flowcharts, program network charts and system resources charts. Geneva: ISO.

ISO 5966:1982. Documentation – presentation of scientific and technical reports. Geneva: ISO.

ISO 7144:1986. Documentation – presentation of theses and similar documents. Geneva: ISO.

ISO 9175-1:1988. Tubular tips for hand-held technical pens using India ink on tracing paper – Part 1: Definitions, dimensions, designation and marking. Geneva: ISO.

ISO 9175-2:1988. Tubular tips for hand-held technical pens using India ink on tracing paper – Part 2: Performance, test parameters and test conditions. Geneva: ISO.

ISO 9178-1:1989. Templates for lettering and symbols – Part 1: General principles and identification markings. Geneva: ISO.

ISO 9178-3:1989. Templates for lettering and symbols – Part 3: Slot widths for technical pens with tubular tips in accordance with ISO 9175-1. Geneva: ISO.

IUB [International Union of Biochemistry]. 1978. Biochemical nomenclature and related documents. London: Biochemical Society. [For further IUPAC-IUB Joint Commission on Biochemical Nomenclature recommendations on nomenclature see issues of the *European Journal of Biochemistry*.]

IUPAC [International Union of Pure and Applied Chemistry]. 1971. Nomenclature of inorganic chemistry, 2nd ed. Oxford: Pergamon Press [Reprinted 1981].

IUPAC [International Union of Pure and Applied Chemistry]. 1979. Nomenclature of organic chemistry. Oxford: Pergamon Press.

Iverson C, Dan B B, Glitman P et al. 1989. American Medical Association manual of style, 8th ed. Baltimore, MD: Williams & Wilkins.

Krieg N R, Holt J G eds. 1984 & 1986. Bergey's manual of systematic bacteriology, 2 vols. Baltimore, MD: Williams & Wilkins.

Mabberley D J. 1987. The plant-book: a portable dictionary of the higher plants. Cambridge: Cambridge University Press.

Marler E E J compiler. 1985. Pharmacological and chemical synonyms: a collection of names of drugs, pesticides and other compounds drawn from the medical literature of the world, 8th ed. Amsterdam: Elsevier.

The Merck index: an encyclopedia of chemicals, drugs, and biologicals, 11th ed. (ed. Budavari S et al.) 1989. Rahway, NJ: Merck.

O'Connor M. 1978. Editing scientific books and journals. Tunbridge Wells: Pitman [now available from the European Association of Science Editors].

O'Connor M. 1986. How to copyedit scientific books & journals. Philadelphia: ISI Press [now available from Williams & Wilkins, Baltimore, MD].

O'Connor M, Woodford F P. 1975. Writing scientific papers in English: an ELSE-Ciba Foundation guide for authors. Amsterdam: Excerpta Medica.

Paine R T, Vadas R L. 1969. The effects of grazing by sea urchins, *Strongylocentrotus* spp., on benthic algal populations. Limnology and Oceanography 14:710–719.

Porcher F H. 1986. Reference practices of biomedical journals: uniform requirements style or not? CBE Views 9(2):30–39.

Reeves M, ed. 1989. Microcomputer graphics for geoscientists: short course notes, vol. 5. Dept of Earth Sciences, Memorial University, St John's, Newfoundland A1B 3X5: GAC Publications. 156 pp. 1 9 PC disks. [CDN$50 + $5 postage, payable to Geological Association of Canada.]

Reif-Lehrer L. 1982. Writing a successful grant application. Boston, MA: Science Books International.

Reynolds J E F, ed. 1989. Martindale: the extra pharmacopoeia, 29th ed. London: Pharmaceutical Press.

Reynolds L, Simmonds D. 1981. Presentation of data in science: publications, slides, posters, overhead projections, tape-slides, television – principles and practices for authors and teachers. The Hague: Martinus Nijhoff. [Available from Kluwer, Dordrecht, The Netherlands.]

Roberts P D. 1987. Plain English: a user's guide. Harmondsworth, UK: Penguin.

Robinson A H, Sale R D, Morrison J L. 1978. Elements of cartography, 4th ed. New York: John Wiley. [Cited by CBE Scientific Illustration Committee 1988, p. 284.]

Rosen F S, Steiner L, Unanue E R. 1989. Macmillan dictionary of immunology. London: Macmillan.

Royal Society. 1975. Quantities, units and symbols. London: Royal Society.

Schmidt C F. 1983. Statistical graphics: design principles and practices. New York: Wiley.

Simmonds D, Reynolds L. 1989. Computer presentation of data in science: a do-it-yourself guide based on the Apple Macintosh for authors and illustrators in the sciences. Dordrecht, The Netherlands: Kluwer.

Skerman V B D, McGowan V, Sneath P H A eds. 1980. Approved lists of bacterial names. International Journal of Systematic Bacteriology 30:225–240.

Squires B P. 1989. Biomedical review articles: what editors want from authors and peer reviewers. Canadian Medical Association Journal 141:195–197.

Steinberg J. 1985. Verbal anaemia. New Society p. 90 (19 July).

Strachan D P. 1989. Hay fever, hygiene, and household size. British Medical Journal 299:1259–1260.

Strunk W, White E B. 1978. The elements of style, 3rd ed. Macmillan, New York.

Swanson E, ed. 1979. Mathematics into type. Providence, RI: American Mathematical Society.

Tufte E. 1983. The visual display of quantitative information. Cheshire, CT: Graphics Press.

Tufte E. 1990. Envisioning information. Cheshire, CT: Graphics Press.

Turk C. 1985. Effective speaking: communicating in speech. London: E & F N Spon.

Turk C, Kirkman J. 1989. Effective writing: improving scientific, technical and business communication, 2nd ed. London: E & F N Spon.

Ulrich's international periodicals directory, 28th ed. 1989. New York: Bowker. 3 vols.

U.S. Geological Survey. 1978. Suggestions to authors of the reports of the United States Geological Survey, 6th ed. Washington, DC: U.S. Government Printing Office.

Walker M J C, Lowe J J. 1982. Lateglacial and early Flandrian chronology of the Isle of Mull, Scotland. Nature (London) 296:558–561.

Webb E C, ed. 1984. Enzyme nomenclature: recommendations (1984) of the Nomenclature Committee of the International Union of Biochemistry. Orlando, FL: Academic Press.

Woodford F P. 1967. Sounder thinking through clearer writing. Science (Washington) 156:743–745.

Woodford F P, ed. 1968. Scientific writing for graduate students: a manual on the teaching of scientific writing. Bethesda, MD: Council of Biology Editors (4th printing, 1986).

Woolsey J D. 1989. Combating poster fatigue: how to use visual grammar and analysis to effect better visual communications. Trends in Neurosciences 12:325–332.

Woolston D C, Robinson P A, Kutzbach G. 1988. Effective writing strategies for engineers and scientists. Chelsea, MI: Lewis Publishers.

Zeiger M. 1991. Essentials of writing biomedical research papers. New York: McGraw-Hill.

FOR FURTHER READING OR REFERENCE

Bénichoux R, Michel J, Pajaud D et al. 1985. Guide pratique de la communication scientifique: comment écrire, comment dire. Paris: Gaston Lachurié.

218

Farr A D. 1985. Science writing for beginners. Oxford: Blackwell Scientific.

Gastel B. 1983. Presenting science to the public. Philadelphia, PA: ISI Press [now available from Williams & Wilkins, Baltimore & London].

Huth E J. 1990. How to write and publish papers in the medical sciences, 2nd ed. Baltimore, MD & London: Williams & Wilkins.

IUB-CEBJ [International Union of Biochemistry-Commission of Editors of Biochemical Journals]. 1973. The citation of bibliographic references in biochemical journals. Recommendations (1971). Biochemical Journal 135:1–3 [and in other biochemical journals].

Lock S. 1985. A difficult balance: editorial peer review in medicine. London: Nuffield Provincial Hospitals Trust [now available from Williams & Wilkins, Baltimore & London].

Morgan P. 1986. An insider's guide for medical authors & editors. Philadelphia, PA: ISI Press [now available from Williams & Wilkins, Baltimore & London].

University of Chicago Press. 1987. Chicago guide to preparing electronic manuscripts. Chicago: University of Chicago Press.

University of Chicago Press. 1982. The Chicago manual of style, 13th ed. Chicago: University of Chicago Press.

Index